The Seed of Eternal Blossom

DNA: Past, Present and Future

1ST EDITION

Mohamed M Haq MD

****

COPYRIGHT

– ALL RIGHTS RESERVED –

All rights reserved under International Copyright Conventions. By payment of the required fees, you have been granted the non-exclusive, non- transferable right to access and read the text of this e-book on-screen. No part of this text may be reproduced, transmitted, down-loaded, decompiled, reverse engineered, or stored in or introduced into any information storage and retrieval system, in any form or by any means, whether electronic or mechanical, now known or hereinafter invented, without the exclusive, express written permission of the author, Mohamed M Haq.

eBook Template Copyright by MBS International

COPYRIGHT BY:

The Seed of Eternal Blossom. DNA: Past, Present and Future – Copyright 9 April, 2019 by the author, Mohamed M Haq

Acknowledgements

I am greatly indebted to my teachers from school thru college and post graduate training, who instilled an everlasting love for seeking knowledge. My parents' expectations and confidence in me was an invisible force to propel me at every stage of my life. The camaraderie of my classmates in medical school, characterized by a cordial competition to excel and unconditional love for each other, played a significant role in shaping my personality. These bonds of friendship evolved into lifelong relationships. My classmate and dear friend, Professor Srinath Reddy, the shining star and pride of our class, who rose to become a Professor at Harvard, was the first one to read my manuscript and encourage me in this endeavor. My childhood friend from school and through Medical School, Dr. Vijay Varma, a voracious reader with extensive knowledge in the field was kind enough to give an early review, providing valuable advice. Abbas Ali Khan , a good friend, a non-physician, critically reviewed the work, providing significant feedback on simplifying technical sections and adding some additional segments. Dr. Vilas Deshpande, an accomplished surgeon, and Dr. Mangal Katkeneni, two of my medical school classmates, critically reviewed the entire manuscript. Their suggestions were very helpful in improving the final version. Dr. Nageshwari Krishnareddy, my classmate, neighbor and dear friend, played a key role in shaping Chapter 3, on Hindu beliefs. Several other friends and colleagues voluntarily and enthusiastically were my beta readers, helping to shape the final product. My brother, Ejaz Haq, assisted me in navigating the publishing process. I gratefully acknowledge my wife's patience and understanding to let me work thru writing and rewriting. My children, Ayman and Nureen, and son-in-law Abid, always eagerly and joyfully assisted me with computer glitches, organizing my material and multiple sundry tasks. I want to thank Kevin Anderson from Kevin Anderson and Associates - Ghost Writing and Editing Services, New York, whose advice I sought to see if the manuscript is worth publishing. His commentary and opinion cemented my desire to proceed with publication. Above all, my greatest debt is to the giants in the field, thru the ages, the most brilliant, daring and indefatigable minds, whose work I have the privilege to present in simple words to the readers.

DEDICATION

Dedicated to
My Parents who gave me my DNA.
My wife Shaheen who filled my life with love
Samir, Nureen and Ayman, who not only carry forth our DNA but have been the joy of our lives.

TABLE OF CONTENTS

Contents

The Seed of Eternal Blossom i
COPYRIGHT ii
   – ALL RIGHTS RESERVED – ii
Acknowledgements iii
DEDICATION iv
TABLE OF CONTENTS v
.... 1
INTRODUCTION 2
CHAPTER 1: THE HANDS OF KHNUM SHAPE THE BODY 6
GREAT HYMN TO KHNUM 6
   SOURCES & REFERENCES – CHAPTER 1 8
CHAPTER 2: THE ANCIENT EGYPTIANS' BELIEFS ABOUT BIRTH AND THE HUMAN BODY 9
   THE ANCIENT EGYPTIANS 9
   VIEW KARNAK TEMPLE IN 3D 9
   THE ANCIENT EGYPTIANS' BELIEFS ABOUT THE HUMAN SOUL 10
   THE ANCIENT EGYPTIANS' BELIEFS ON BIRTH 10
   THE ANCIENT EGYPTIANS' BELIEFS ON LIFE AFTER DEATH 10
   IN CONCLUSION 11
   SOURCES & REFERENCES – CHAPTER 2 11
CHAPTER 3: ANCIENT HINDU BELIEFS ABOUT BIRTH AND THE HUMAN BODY 12
HINDUISM 12
   BELIEFS ABOUT THE TRANSMISSION OF CHARACTER AND THE SOUL .. 13
Kunti's Mantra 13
   VIEW THE MAHABHARATA EPIC HERE 13
   MARRIAGE AND INHERITED CHARACTERISTICS 14
   THE HINDUS' BELIEFS ON BIRTH 14
   THE HINDUS' BELIEFS ON LIFE AFTER DEATH 15
   IN CONCLUSION 15
   SOURCES & REFERENCES – CHAPTER 3 15

CHAPTER 4: ANCIENT JUDEO-CHRISTIAN BELIEFS ABOUT BIRTH AND THE HUMAN BODY 16
    ANCIENT JUDEO-CHRISTIAN BELIEFS 16
    JUDEO-CHRISTIAN BELIEFS ABOUT THE HUMAN SOUL 16
    JACOB'S STORY – HIS KNOWLEDGE OF GENETICS 18
    DID JACOB USE GENETICS TO CREATE THE TYPE OF GOATS & SHEEP HE WANTED? 19
    ANCIENT JUDEO-CHRISTIAN BELIEFS ABOUT BIRTH 20
    ANCIENT JUDEO-CHRISTIAN BELIEFS ON LIFE AFTER DEATH 20
    WHAT HAPPENED TO THE 'WICKED'? 21
    IN CONCLUSION 21
    Sources & References – Chapter 4 21

CHAPTER 5: ISLAM'S BELIEFS ABOUT BIRTH AND THE HUMAN BODY 22
    FORMATION of A BABY 22
    ISLAMIC BELIEFS ABOUT THE HUMAN SOUL 23
    ISLAMIC BELIEFS ON BIRTH 24
    ISLAMIC BELIEFS ON LIFE AFTER DEATH 24
    IN CONCLUSION 25
    SOURCES & REFERENCES – CHAPTER 5 25

CHAPTER 6: ANCIENT CHINESE BELIEFS ABOUT BIRTH AND THE HUMAN BODY 26
    THE ANCIENT CHINESE 26
    ANCIENT CHINESE BELIEFS ABOUT THE HUMAN SOUL 27
    ANCIENT CHINESE BELIEFS ON BIRTH 27
    A WOMAN'S INFLUENCE ON AN UNBORN CHILD 27
    ANCIENT CHINESE BELIEFS ON LIFE AFTER DEATH 28
    IN CONCLUSION 28
    sources & REFERENCES – chapter 6 28

CHAPTER 7: LIKE FATHER, LIKE SON - ARISTOTLE, THE GREEKS, AND THE DAWN OF RATIONAL THINKING 30
    LOGOS VS. MYTHOS 30
    WHAT HAPPENED IN GREECE AND HOW IT BECAME THE CENTER OF WISDOM AND KNOWLEDGE 30
    ARISTOCRATIC RULE WAS REPLACED FOR THE FIRST TIME 30
    THE GREEKS RELIED ON OBSERVATION AND ANALOGICAL DEDUCTIONS 31
    SOCRATES AND PLATO – THE BEGINNING OF 'REASON' IN GREECE 31

- ARISTOTLE – INVESTIGATING THE GREATEST PHILOSOPHER OF ALL TIMES: HIS BELIEFS & THOUGHTS ABOUT GENETICS. 32
  - WHO INFLUENCED ARISTOTLE? 32
  - 624 BC – 546 BC: THALES – EVERYTHING IS DERIVED FROM WATER......32
  - 610 BC – 546 BC: ANAXIMANDER – AIR AS THE PRIMORDIAL ELEMENT ..32
  - 535 BC – 475 BC: HERACLITUS – EVERYTHING IS CHANGING. 33
  - 492 BC – 432 BC: EMPEDOCLES – FORCES OF ATTRACTION AND REPULSION GIVE BIRTH TO LIFE..33
  - 460 BC – 370 BC: THE BEGINNING OF THE ATOMIST SCHOOL – EVERYTHING IS MADE OF 'ATOMS' 33
  - 342 BC – 271 BC: EPICURUS – THE CHAMPION OF GREEK MATERIALISM 33
  - 460 BC – 370 BC: HIPPOCRATES – THE FORMATION OF THE EMBRYO AND HEREDITY 34
  - BACK TO ARISTOTLE AND HIS UNDERSTANDING OF INHERITANCE 34
  - THE FORMATION OF AN EMBRYO 35
  - ROMAN RULE AND GREEK LEGACY 35
  - GALEN – THE EMPEROR OF MEDICINE 36
  - IN CONCLUSION 36
  - SOURCES & REFERENCES – CHAPTER 7 36
- CHAPTER 8 THE RETURN OF MYTH & MYTHOLOGY; THE LONG DARK AGES 37
  - THE PERCEPTION OF BIRTH, GENETICS AND HEREDITY DURING THE LONG DARK AGES..37
  - CICERO'S PURSUIT OF LEARNING & EVALUATING THE ROMAN EMPIRE'S ATTITUDE TO NATURAL SCIENCES, EVEN BEFORE THE DARK AGES 38
  - THE ROMAN VIEWS ON NATURAL SCIENCE STUDIES 39
  - THE CONCEPT OF HEREDITY IN THE ROMAN EMPIRE AND BEFORE THE LONG DARK AGES..39
  - 5TH – 15TH CENTURY AD - CHANGES IN THE CONCEPT OF HEREDITY DURING THE LONG DARK AGES 40
  - THE CONCEPT OF 'ENSOULMENT' 40
  - INVESTIGATING THE CHANGING IDEAS ABOUT 'ENSOULMENT' 41
  - 14TH THRU 17TH CENTURIES – THE RENAISSANCE 41
  - IN CONCLUSION..41
  - SOURCES & REFERENCES – CHAPTER 8 41
- CHAPTER 9: THE LIGHT AT THE END OF A DARK TUNNEL 42
  - PLANT REPRODUCTION AND TRANSMISSION OF CHARACTERISTICS. ...42

WILLIAM HARVEY'S THEORY OF EPIGENESIS AND THE CHANGES OF THE CONCEPT OF HEREDITY AFTER THE DARK AGES 43

    IN THE SHADOWS 43

    IN CONCLUSION 44

    SOURCES & REFERENCES – CHAPTER 9 44

CHAPTER 10: MAN - CREATED IN GOD'S IMAGE OR A BRANCH OF THE MAJESTIC TREE OF LIFE? 45

    CARL UNDERSTOOD SEXUAL REPRODUCTION IN PLANTS FROM A YOUNG AGE 45

    CARL AT THE UNIVERSITY OF LUND 46

    CARL LINNAEUS AT UPPSALA UNIVERSITY – THE DEVELOPMENT OF THE BINOMIAL SYSTEM 46

    UNDERSTANDING WHAT THE BINOMIAL SYSTEM IS 46

    THE LEGACY OF CARL LINNAEUS 47

    IN CONCLUSION 47

    SOURCES & REFERENCES – CHAPTER 10 47

CHAPTER 11: THE INVISIBLE COMES TO LIFE 48

    LIVING ANIMALCULES THROUGH ANTONI'S MICROSCOPE 48

    THE HISTORY OF MICROSCOPES 48

    LEEUWENHOEK'S EARLY LIFE AND HIS JOURNEY TO BECOME THE GREATEST LENS-MAKER 49

    ANTONI LEEUWENHOEK'S MICROSCOPES 49

    LEEUWENHOEK'S CORRESPONDENCE WITH THE ROYAL SOCIETY 50

    THE THEORY OF SPONTANEOUS GENERATION AND LEEUWENHOEK'S LEGACY 50

    AN UNFORTUNATE LACK OF INTEREST AFTER VAN LEEUWENHOEK'S DEATH 51

    IN CONCLUSION 51

    SOURCES & REFERENCES – CHAPTER 11 51

CHAPTER 12: LIFE BEGETS LIFE: CELL THEORY 52

EXISTING VIEWS ABOUT THE FORMATION OF LIVING THINGS 52

    SAINT AUGUSTINE OF HIPPO'S VIEW OF DIVINE DECREE 52

    ARISTOTLE'S THEORY OF SPONTANEOUS GENERATION – LIVING THINGS COULD COME FROM NON-LIVING THINGS 53

    REDI'S FAMOUS EXPERIMENTS - THE FIRST EVIDENCE AGAINST SPONTANEOUS GENERATION 53

THE INVENTION OF THE MICROSCOPE AND THE ORIGINS OF CELL THEORY 54
    SCHLEIDEN AND SCHWAN 54
    SCHWAN AND SCHLEIDEN WERE NOT THE FIRST ONES TO TALK ABOUT 'CELLS' AS FUNDAMENTAL BUILDING BLOCKS IN LIVING ORGANISMS 55
    WHO WAS RENE DUTROCHET? 55
    DUTROCHET'S WORK AND HIS CONTRIBUTION TO PLANT PHYSIOLOGY 55
    MODERN CELL THEORY AND VIRCHOW'S ARCHIVES 56
    IN CONCLUSION 56
    SOURCES & REFERENCES – CHAPTER 12 56
CHAPTER 13: IF MAN CAN DO IT, WHY NOT NATURE? DARWIN: EVOLUTION BY NATURAL SELECTION 57
    CHARLES DARWIN 57
    THE MOST IMPORTANT SCIENTIFIC VOYAGE IN HISTORY 57
    DARWIN'S FINCHES 58
    THE PROCESS OF NATURAL SELECTION 58
    CHARLES DARWIN AND ALFRED WALLACE 59
    PANGENESIS AND GEMMULES 59
    PROOF IS FOUND 60
    IN CONCLUSION 60
    sources & REFERENCES – CHAPTER 13 60
CHAPTER 14: MONASTERY TURNED SCIENCE LAB: MENDEL, THE FATHER OF GENETICS 61
    THE LIFE AND DEATH OF JOHANN MENDEL 60
    MENDEL AT THE UNIVERSITY OF VIENNA 62
    WHILE AT THE MONASTERY 62
    THE BEGINNING OF MENDEL'S RESEARCH 62
    MENDEL'S PEAS 63
    THE MENDELIAN LAWS OF HEREDITY 64
    IN CONCLUSION 64
    sources & REFERENCES – CHAPTER 14 65
CHAPTER 15: LORD OF THE FLIES: FRUIT FLIES FOLLOW THE LORD'S COMMAND 66
    THOMAS HUNT MORGAN 66
    THE FLY AND THE LITTLE ROOM 67
    CHROMOSOMES X AND Y 68

    THE FLY PEOPLE 68
    IN CONCLUSION 69
    SOURCES & REFERENCES – CHAPTER 15 70

CHAPTER 16: NOBEL INJUSTICE: AVERY ESTABLISHES DNA AS THE CARRIER OF HEREDITY 71

    EARLY LIFE – AVERY WAS DESTINED TO BECOME A CLERGYMAN BUT ENDED UP A PHYSICIAN 71
    AVERY AT ROCKEFELLER INSTITUTE AND THE CURE FOR PNEUMONIA 72
    WORLD WAR I AND AVERY'S RESEARCH ON INFLUENZA 72
    AVERY DISCOVERS A NEW METHOD TO IDENTIFY THE TYPE OF PNEUMOCOCCUS BACTERIA 72
    THE DISCOVERY OF 'SUGARCOATED MICROBES' 72
    AVERY AND RENE DUBOS – THE DISCOVERY OF THE FUNDAMENTAL UNIT OF HEREDITY 73
    FREDERICK GRIFFITH'S DISCOVERY OF STRAINS 73
    THE 'TRANSFORMING PRINCIPLE' – AVERY REJECTS CONCLUSIONS 73
    AVERY'S CONTRIBUTION TO THE 'TRANSFORMING PRINCIPLE' AND HIS DISCOVERY OF DNA AS THE GENE CARRIER 74
    IN CONCLUSION 75
    Sources & References: chapter 16 75

CHAPTER 17: IMAGINATION OVER EXPERIMENTATION; WATSON AND CRICK - THE DNA MODEL 76

    SCIENTISTS WHO CONTRIBUTED TO WATSON AND CRICK'S DNA MODEL – THE GROUNDWORK 76
    FRIEDRICH MIESCHER 76
    PHOEBUS LEVENE 77
    ERWIN CHARGAFF 77
    LINUS PAULING 78
    JAMES WATSON AND THE DISCOVERY OF DNA STRUCTURE 79
    WATSON MEETS CRICK AT CAMBRIDGE UNIVERSITY 79
    THE DOUBLE HELIX MODEL OF DNA 79
    THE NEW AND IMPROVED MODEL 80
    IN CONCLUSION 80
    SOURCES & REFERENCES – CHAPTER 17 80

CHAPTER 18: DNA SLEUTHS: CRACKING THE CODE 81
    GAMOW 81

DNA AND RNA 81
THE RNA TIE CLUB 82
KHORANA TAKES SCIENCE TO THE HOLY GRAIL 82
NIRENBERG 82
NIRENBERG AND MATHAEI'S EUREKA MOMENT 83
NEWS REPORT 84
THE RACE FOR THE CODE 84
IN CONCLUSION 85
References: 85

CHAPTER 19: DNA GOES TO WORK: ONE GENE, ONE PROTEIN 86
ARCHIBALD GARROD 86
BEADLE AND TATUM 87
NEUROSPORA 87
IN CONCLUSION 88
sources & References – chapter 19 88

CHAPTER 20: RECOMBINANT DNA: TAKING ARTIFICIAL BREEDING TO THE NEXT LEVEL 89
THE MEETING 89
BACTERIA ARE SINGLE CELL CREATURES 90
BACTERIA AND THE CHROMOSOMES 91
PAUL BERG 91
COHEN AND BOYER 92
IN CONCLUSION 93
sources & REFERENCES – CHAPTER 20 93

CHAPTER 21: BACTERIA & YEAST: THE NEW INDUSTRIAL WORKER 94
CREATING HUMULIN 95
VITAMINS 96
RICE 96
IN CONCLUSION 96
SOURCES & REFERENCES – CHAPTER 21 97

CHAPTER 22: GAIN BECOMES PAIN AS THE CLOCK TURNS 98
DOCTOR TONY ALLISON 98
SICKLE CELL DISEASE 99
SICKLE CELL AND NATURAL SELECTION 100
V LEIDEN MUTATION AND NATURAL SELECTION 101

IN CONCLUSION 101

sources & REFERENCES – CHAPTER 22 102

CHAPTER 23: UNJUST INCARCERATION 103

THE STORY 103

THE TRAIL 104

WHAT HAPPENED NEXT 104

THE DNA TEST 105

DNA PROFILING: ALEC JEFFRIES' INNOVATION......................................106

IN CONCLUSION 107

sources & REFERENCES – CHAPTER 23 108

CHAPTER 24: ULTIMATE GENEALOGY: TRACING THE ROOTS OF ADAM & EVE 109

THOR HEYERDAHL AND THE CURRENTS 109

DNA AND ANTHROPOLOGY 110

THE ICEMAN OF THE OTZAL MOUNTAINS 110

BRYAN SYKES AND OTZI 111

BUT WHAT ABOUT THE MEN? 112

HAS SCIENCE TRACED ADAM AND EVE? 112

CONCLUSION 112

Sources and References – Chapter 24 112

CHAPTER 25: SEARCHING FOR LUCA (LAST UNIVERSAL COMMON ANCESTOR) IN THE RNA WORLD..113

LUCA...113

THE FIRST MOLECULES OF LIFE 114

IN CONCLUSION 115

Sources and References – chapter 25..115

CHAPTER 26: ORPHANS IN A RICH KINGDOM: VIRUSES................................116

SMALLPOX AND COWPOX 116

TOBACCO VIRUSES..117

THE FIRST VIRUS..117

THE GOOD VIRUS 118

WAIT, WHAT IS LIFE? 118

SO, VIRUSES? 118

LIFE ON MARS 119

IN CONCLUSION 119

Sources and References – chapter 26..120

CHAPTER 27: QUEST FOR SUPERMAN: THE GOOD, BAD AND THE UGLY 121
- THE CASE 121
- EUGENICS 121
- EUGENICS IN ENGLAND 122
- THE NAZIS' EUGENICS 122
- BERG'S RESEARCH 123
- IN CONCLUSION 125
- sources & REFERENCES – CHAPTER 27 125

CHAPTER 28: MINDS SHAPE THE FUTURE 126
- SWEET 16 AND NEVER BEEN HEALTHIER 126

VIOLENCE AS A RESEARCH SUBJECT 128
- MONOAMINE OXIDASE 129

HOW ABOUT CLONING OUR ORGANS OR OUR WHOLE BODY AS SPARE PARTS? 130
- CAN SCIENCE FIX DEFECTIVE GENES? 131

THE FUTURE OF DNA AND SCIENCE LOOMS BRIGHT FOR YOUNG MINDS   131

IN CONCLUSION 132
- REFERENCES AND SOURCES: CHAPTER 28 133

POST SCRIPT 134

WOW! WHAT A MARVELOUS JOURNEY 134

HUMAN CELL STRUCTURE 135

FURTHER READING 136

ABOUT THE AUTHOR 137

GLOSSARY 138

Could the knowledge encoded deeply in the cells of our bodies hold the key to liberation from disease, mental issues and destruction for our own society?

INTRODUCTION

Jean-Baptiste Tavernier, a French traveler in the mid-17th century, visited Hyderabad; the city where I grew up, in India. He was so impressed by the city's beauty and grandeur that he felt it matched his own Paris, considered the most beautiful city in Europe at the time.

It was also the region which produced some of the largest diamonds in the world, including the *Kohinoor*, (which adorns the British crown today), and *Hope*, the pride of the Smithsonian gem collection in Washington DC.

Dr. Ronald Ross, a British physician conducting research in my hometown, discovered the link between the dreaded killer, Malaria, and mosquitoes. As a result, he was awarded the coveted Nobel Prize for medicine in 1902.

The hallowed ground on which stood the hospital where I met my first patient was known for its Anesthesia faculty in the 1880's; it had devised a safe method of administering anesthesia by chloroform. The method, named the 'Hyderabad Cap,' was adopted by Great Britain and the west.

During my second year studying medicine, in 1968 at the Osmania Medical College in Hyderabad, India, one of the topics covered in the syllabus was genetics. It was a small section, as knowledge in the field was elementary and its application in clinical medicine limited. Thankfully, the college library acquired a book on genetics that year. It was written in a different style, unlike the typical textbooks of medicine. It emphasized a promise of the future, rather than reporting the available knowledge. This book changed my life and perspective in many ways, and I became more and more interested in learning as much as I could about genetics and its effect on human life.

My enthusiasm was further piqued when, in October that same year, the entire population of India was swept up in a wave of excitement and pride as radio and newspapers announced that Har Gobind Khorana, a US-based scientist of Indian descent, had won the Nobel Prize for Physiology or Medicine. Alongside two others, Marshall W. Nirenberg and Robert W. Holley, Khorana was nominated for his innovative and elegant technique combining enzymatic and synthetic methodology, to create nucleotides from simple chemicals in a test tube. Nucleotides are the building blocks of DNA and RNA, the molecules of life. I was hooked on genetics from that moment onwards. It became my favorite subject. I followed discoveries in the field with immense interest and enthusiasm throughout my life. Not surprisingly, genetics became one of the most important fields in medicine in the twentieth century, as scientists discovered it held the formula for unraveling the secrets of life.

Genetics appeared to answer some of the fundamental questions about life:

What is life chemically made of?
How do the cells and organs in the body work?
How is it that two separate individuals produce a new life form that contains features of both and of the generations before them?
What was it that determined the array of different skin colors in our world?
Why are some children born with anomalies?
Why do cats give birth to kittens, and not puppies?

How can a few simple chemicals produce such a panorama of life forms on our planet?

The perpetual pursuit of unraveling the hidden treasures of nature seemed to be reaching the Holy Grail. Questions were many, answers few.

After completing my medical degree, I migrated to the United States in 1975 to pursue advanced training in *Internal Medicine and Medical Oncology* at the renowned MD Anderson Cancer Center.

By then, science had progressed on a relentless path. Not only had we begun deciphering the secrets of life hidden in our DNA, but we were also learning how to manipulate it for many uses to benefit human society; from the manufacture of biological molecules for medicinal use, to applications in forensic investigations. The chemical and molecular basis of some of the diseases plaguing humankind had been deciphered, opening specific and effective treatments for these illnesses.

As discoveries mounted and we learnt more, it became clear that we had only touched the tip of the iceberg. As with many developments in science, especially those pertaining to the secrets of life, more questions arose. At first, the very building blocks seemed simple and common to all living structures. However, scientists soon discovered that nothing in our planet or universe is 'simple,' but all is immensely complex. As Einstein had remarked, *"the more I learn, the more I realize how much I don't know."* Nevertheless, discoveries in biology seemed to be catching up with the spectacular advances in other fields, such as electronics, telecommunications and machines. As for what we know today, we can be rightly proud of the progress made, but ever aspiring to know more in order to assist humanity and society in our quest to make life better for ourselves.

Unfortunately, people's perception of science is often influenced less by formal or informal education than by popular culture. Therefore, it is rather difficult for a science article or a book to compete with a movie such as *Jurassic Park, Harry Potter* or *Marvel's* many superhero fantasies. Mainstream popular culture often bends the laws of nature and the advances of science to suit its story-line and heighten the entertainment level for its viewers. Albeit such entertainment is short-lived and only lasts an hour or two, viewers are influenced to some degree, by what they see.

Max, the transgenic lead female character of the TV series, *Dark Angel 2000*, has genes of both a human and a cat, giving her an exceptional sense of smell, while another character looks like a man, but has the head and hands of an iguana. Due to the entertaining storylines, plot and characters, the attention of fans worldwide is piqued. This is not such a 'bad thing,' as science does get some free publicity in a way, although the 'science' or genetics presented is often not accurate or in line with the latest research.

Nevertheless, scientists and academic writers can learn quite a bit from the most popular entertainment out there. Science and scientists are, of course, neither romantic nor prone to creating dramatic plotlines; they are more like humorless drones, focused in earnest on finding the answers to the multitude of questions driving them forward each day. It would help to learn how to covey science in a more consumer-friendly manner, so that we can spread the word about the new and exciting findings we make.

The goal of medicine is to enhance the quantity and quality of life. Doctors use medicine, surgery and other interventions to achieve these two goals. Through the years, in my interaction with patients, I have found that, when I take the time to explain the nature of illness, our knowledge in the field, its limitations and how the knowledge is being channeled to help their individual case, the information fascinates them, gives them comfort and, more importantly, makes them a part of the treatment team with the nurses and other staff. Most often, patients have very limited knowledge, and the knowledge they have was shaped by popular culture, and at times, mixed with inaccurate information, especially about genetics, DNA and its applications. Hyped media articles, and blogs by novices who overheard a concept, confuse and at times mislead people. Even my medical colleagues, whose daily routine did not require significant knowledge in the area of genetics, had very limited awareness of recent discoveries. Whenever I spoke about genetics and DNA to my patients, friends, colleagues and family, both about the findings of modern science and about ancient science, they were immensely intrigued and wanted to know more.

This prompted me to write a review article on the subject in 1993. *Texas Medicine,* a journal that reaches most of the practicing doctors in the state of Texas, published it. To my surprise, the article was very well received and recognized with the *"Harriet Cunningham Citation"* for meritorious writing of a scientific review article.

Stories about the evolution of knowledge in the field of heredity, genetics and DNA are fascinating and riveting. When I shared these stories of the men and women, the pioneers who dared to think outside the box, with friends, family and colleagues, they would say: "Oh, my God! I did not know that." I soon realized that most of modern humanity didn't know the secrets of their past or the knowledge left behind by our ancestors. I believe it is not simply curiosity, but it is necessary for society to know these facts that will help us, as human beings, to embrace our past and look to the future, knowing that there is a solution to our problems. The intertwined coil of DNA is an iconic figure known to everyone, but, as symbolic as it is, it not only has meaning to those in medicine, but it can have a tremendous effect on everyone, if they understand its secrets, and learn how to apply them in their daily lives.

DNA is powerful! We know already that it can be used in many ways. DNA matching has revolutionized criminal investigations and helped in determining paternity. It has improved farming, diagnosing diseases, developing treatments, producing medicines and many other areas. However, just as the knowledge of atoms can be used for making electricity as well as bombs, knowledge of DNA is beneficial but has the potential for misuse. Individuals from around the world need to know the pros and cons, as research in this field affects each and every one of us. Society's knowledge and participation in public policy can guide funding for appropriate research and regulations can be developed to prevent misuse.

My quest and aspirations are to share this exciting tale of life in a series of short stories, each presented in a simple and easy to follow style. Each chapter highlights a specific area of genetics and the scientist(s) associated with it. In general, each chapter is laid out in chronological sequence, with one dovetailing into the next. However, one can also read a single chapter and still gain insight.

The book covers the stories of some of the known and unknown ancient scientists who made amazingly astute observations, followed by giants in the field over the years, on whose shoulders present day scientists stand to accomplish further breakthroughs. I also speculate a little about the future. My earnest desire is to disseminate knowledge about life so that people will have more information to take better care of themselves and our society at large. I also hope to inspire students, young and old, to pursue studies and research in the field of heredity and genetics, to create a brighter future for humanity and the world.

Come with me on this journey of discovery, and enjoy the knowledge of ancient and modern science in the quest to understand life.

CHAPTER 1:
THE HANDS OF KHNUM SHAPE THE BODY

"How could it be, that an ordinary couple huffing and puffing in the dark, create a new being? God is the only one with the power to create life."

- Edward Dolnick

The ancient Egyptians believed that the Sun God, Ra, also known as Khnum, created each baby with his own hands, and placed it in the mother's womb.

500 miles south of Cairo, on the banks of the Nile, in the town of Esna, is a partially excavated temple dedicated to Ra. On the walls, a beautiful hymn in hieroglyphics describes in meticulous detail the making of a baby. Ra takes the mud and water of the Nile to fashion the fetus.

GREAT HYMN TO KHNUM

God of the potter's wheel,
Who settled the land by his handiwork;
Who joins in secret,
Who builds soundly,
Who nourishes the nestlings by the breath of his mouth;
Who drenches this land with Sun,
While round sea and great ocean surround him.

He has fashioned gods and men,
He has formed flocks and herds;
He made birds as well as fishes,
He created bulls, engendered cows.

He knotted the flow of blood to the bones,
Formed in his workshop
He makes women give birth when the womb is ready,
So as to open as he wishes;
He soothes suffering by his will,
Relieves throats, lets everyone breathe,
To give life to the young in the womb.

He made hair sprout and tresses grow,
Fastened the skin over the limbs;
He built the skull, formed the cheeks,
To furnish shape to the image.

He opened the eyes, hollowed the ears,
He made the body inhale air;
He formed the mouth for eating,
Made the gorge for swallowing.

He also formed the tongue to speak,
The jaws to open, the gullet to drink
The throat to swallow and spit.

The spine to give support,
The testicles to move,
The arm to act with vigor,
The rear to perform its task.

The gullet to devour,
Hands and their fingers to do their work,
The heart to lead.
The loins to support the phallus,
In the act of begetting.
The frontal organs to consume things,
The rear to aerate the entrails,
Likewise to sit at ease,
And sustain the entrails at night.
The male member to beget,
The womb to conceive,
And increase generations in Egypt.
The bladder to make water,
The virile member to eject,
When it swells between the thighs.
The shins to step,
The legs to tread,
Their bones doing their task,
By the will of his heart.

Who among us is not inspired by the phenomenon that was the Egyptian civilization?

Colossal monuments, mummies, tombs, gigantic pyramids, mysterious beliefs, and enormous temples built by almost super-human power, fill us with a sense of wonder and excitement. Even more jaw dropping, though, was the fact that this nation state (the largest of its kind in those days) grew from humble beginnings around 7000 BC, when Egypt was anything but the desert it is today. Geo-archaeologists argue that, from about 8500 BC to 3550 BC, the Sahara was a green savanna that supported a variety of life forms. The Nile was considered the lifeblood of the land by ancient Egyptians, who were as dependent on it then as their modern counterparts, since the river's annual floods provided the water and soil necessary for agriculture.

Ancient Egypt was more than pyramids, temples, and mummies. It was a civilization that gave the world the tools, the technology and practically everything that shaped the societies to follow, even to this day. Their contributions are too numerous to elaborate here. Their inventions include paper, the introduction of alphabets, mathematics, the making of simple machines, cosmetics, a calendar, time keeping devices, construction, an organized labor system, ship-building technology using aerodynamics, an irrigation system using hydraulics, ox-drawn plows, weapons for hunting and warfare, crafts including glass and furniture, to name but a few.

This was the first human civilization that explored and studied, in meticulous detail, the phenomena of nature. It had the scientific spirit of inquiry, the intellectual capacity to innovate, organize and record observations and, more importantly, utilize them to improve society. The Egyptians applied their knowledge and technologies to study the human body. Their belief in the after-life and the practice of mummification to preserve the body provided them with a unique opportunity to examine the human body in great detail. They also devised tools and techniques for surgery and had treatments for many ailments. The hymn to Khnum depicts the depth of their knowledge about the various human organs.

The University of Copenhagen, Denmark has the largest collection of ancient Egyptian papyrus manuscripts on medicine, many of them still in their original language. An international team of scholars is busy at work to translate and publish them. The first conference of the collaboration group to study these scientific papyri was held on May 31st, 2018 at the University of Copenhagen. One can only imagine what the Egyptians could have accomplished if they had had access to modern tools and technology and had not been hampered by religious dogmas.

SOURCES & REFERENCES – CHAPTER 1

Brewer, Douglas J., and Emily Teeter. *Egypt and the Egyptians*. Second edition. Cambridge University Press, 2007.

CHAPTER 2:
THE ANCIENT EGYPTIANS' BELIEFS ABOUT BIRTH AND THE HUMAN BODY

Tutankhamun speaks: "Inside my mask of gold, lapis, turquoise and amber, my naked body breaks its bondage and soars like an eagle towards the golden Sun of rejuvenation." - Ramon Ravenswood

As modern human beings, can we learn anything from our ancient ancestors? Let's find out:

THE ANCIENT EGYPTIANS

Some of the largest and most researched pyramids can be found in Egypt. What is often overlooked, though, are Egyptian temples; an error, probably because of the emphasis given by ancient travelers to the impressiveness of the pyramids, or perhaps because Christian travelers placed little value on 'foreign-idols' or 'barbaric' religions. Some even refused to visit ancient temples, claiming they were places of evil. Others, especially militants, turned to destruction of the beliefs and ancient works of other civilizations.

Yet, Egyptian temples are some of the most phenomenal creations by any ancient civilization. As a matter of fact, some of the oldest and largest places of religion can be found in Egypt, inscribed with scripture upon scripture covering the creation, birth and development of the human body.

The Karnak Temple (*Ipet-Sun*) for instance, is the largest temple or religious building ever constructed. Even modern temples or places of religion cannot compete with the immenseness of the project.

Covering well over 100 hectares, Karnak, also known as the 'Most Sacred of Places,' was dedicated to a variety of select Egyptian gods. Some of them were the god Ra, his wife, the goddess Mut, Munto, the god of war, represented by a falcon-head, and Aten, the sun disk.

VIEW KARNAK TEMPLE IN 3D

To experience the immenseness of Karnak Temple and the majority of other temples built by the ancient Egyptians, a project known as the **UCLA Digital Karnak Project**, reconstructed the temple in 3D:

Video 1: https://www.youtube.com/watch?v=QXDipKxgU24
Video 2: https://www.youtube.com/watch?v=ROg1-enJEms

THE ANCIENT EGYPTIANS' BELIEFS ABOUT THE HUMAN SOUL

What inspired ancient Egyptians to take on projects such as these; for some of them, knowing that the construction might take well over 2,000 years?

How did they even come to imagine the creation of these temples, let alone develop the technology to construct them? A feat that continues to baffle modern-day scientists, engineers and archaeologists.

Furthermore, what immense belief and trust in gods moved the ancient people to completely and utterly dedicate their whole life to their gods, with body, mind and soul?

In order to find answers to these questions, we need to delve deeper into the belief system and knowledge of the ancient Egyptians.

To the ancient Egyptians, a human's *Ka* was part of his/her spirit, also known as the *life source*. (Similar to the Judeo-Christian belief in the concept of the 'soul'.)

According to the Egyptian religion, the soul split into two parts after death: *Ba* and *Ka*. Every morning, *Ba* left the tomb to watch over the deceased's living family. *Ka* spent the day in the heavenly land. At night, *Ba* and *Ka* returned to the tomb to rest until the next sunrise.

THE ANCIENT EGYPTIANS' BELIEFS ON BIRTH

To the Egyptians, the beginning of life was a spiritual, rather than biological process. They believed that the god, Khnum, created the bodies of children from clay on a potter's wheel, and then inserted them into the mother's womb. At the moment of birth, the goddess Meskhenet, (also called Heket in some parts of Egypt), breathed *Ka* into the child's body, giving it life.

In addition to the physical, the Egyptians believed in a metaphysical component of the human body. For example, in addition to the physical heart, there was also a spiritual heart, (or *ib),* that was formed from a single drop of blood from the mother's heart at conception. Ancient Egyptians believed *ib*, and not the brain or mind, was responsible for emotion, thought, determination, and intention. The *ib* survived into the afterlife, where it revealed the character of the person - good or bad.

What is even more interesting is the Egyptian belief that there was a link between the parent and child, with continuity in both physical and spiritual traits. (Pretty much as we, as modern humans, believe that our children will inherit certain physical traits from us through our DNA.)

THE ANCIENT EGYPTIANS' BELIEFS ON LIFE AFTER DEATH

The belief in the transmigration of souls was prevalent in Egyptian mythology, as in some other ancient cultures. The soul, or the spirit, passed from one being - human or otherwise - to another, as the soul did not die. It simply left the physical body of the deceased to enter a new body, which was occupied at birth.

IN CONCLUSION

When we look at the ancient Egyptians, we come to know through the avid recording of their history via religious writings, poems, songs, hieroglyphics, and oral histories that they continually sought to research and investigate the issues of life, afterlife, man's search for righteousness and moral high ground.

The belief and understanding of inheritance derived from their scriptures were further supplemented by mythology, religious dogmas, superstition, and philosophical speculation, some of which continue to influence human thinking in modern times. The ancient Egyptians can provide us with useful insight into the knowledge and wisdom they possessed, by studying the subjects they explored, described and developed, to see how methodically and systematically they attempted to learn about human body. Such devotion to life cannot go unnoticed. Modern science today can glean a whole lot of insight from the study of ancient sciences and will become much the wiser for it.

SOURCES & REFERENCES – CHAPTER 2

Fletcher, Joann. *The Egyptian Book of Living and Dying: The Illustrated Guide to Ancient Egyptian Wisdom*. Watkins Media Limited, 2009.

CHAPTER 3:
ANCIENT HINDU BELIEFS ABOUT BIRTH AND THE HUMAN BODY

"For the soul there is never birth or death, nor, having once been, does it ever cease to be: It is unborn, eternal, everlasting, undying and primeval. It is not slain when the body is slain." - Bhagavad Gita

Were the Egyptians the only ancient civilization who had knowledge of the human body and inheritance of traits?
Let's find out:

HINDUISM

Three thousand miles to the east of the Nile, in what is modern-day India, another civilization flourished, starting around 3500 BC. It did not leave any pyramids, papyruses or hieroglyphics behind; however, it left a vast collection of traditions and beliefs, which were first handed down orally, and subsequently preserved in written form.

The religious scriptures of the Hindus, called the Vedas, were not the teachings of a single messiah, but they provide a rich glimpse into the knowledge and wisdom of the time. In the absence of a single book or an authoritative figure like a pope, many interpretations and practices characterize the diversity of Hinduism.

India was not just a land of sages preaching spirituality and expounding the nature of the soul and the creator, but also a place which bred mathematicians, astronomers, physicians, surgeons and metallurgists. Aryabhata started the decimal system. The notation system now known as Arabic numerals originated in India and was only spread to the rest of the world by the Arabs. The idea of a binary system, the core of computer languages, was described in a treatise on the rules of poetic meters and verses. Brahmagupta, in the seventh century, laid out the rules governing the use of zero, of negative numbers, and formulating equations. Surgical techniques for cataract and plastic repair of the nose were detailed, besides a system of medicine known as Ayurveda, which is practiced even today. India also achieved success in the science of metallurgy, smelting zinc and later producing the first seamless metal globes during the reign of the Mughal emperors in the 16th century. Yoga and meditation are other timeless gifts of ancient India with increasing popularity all across the world to this day.

BELIEFS ABOUT THE TRANSMISSION OF CHARACTER AND THE SOUL

Kunti's Mantra

*"vishnur yonim kalpayatu twashtur roopnai vigam shatu
aasimprajantu prajapatirdhata gharbham vidha dhatu
garbham dehi cinivaari garbham dehi saraswati garbhante
ashwni-ou devi pradhattam pushpa raja stree sa."*

Kunti, a princess in ancient Indian mythology, was given these verses, a mantra, by a sage who was impressed by her hospitality. Invoking this mantra or 'charm' had the power to allow Kunti to become pregnant with a son from a god of her choice. The child would then be born with the features, power and character of the god.

Kunti's mantra in Sanskrit, like the epic, *Mahabharata*, is based on oral stories and legends preserved among the tribes of northern India. The epic, one of the oldest Indian classics, revered for its deep spiritual wisdom, tells the story of King Pandu's five heroic sons, whose destiny was to rule a vast kingdom by defeating the one hundred sons of their blind uncle, King Dhritarashtra. The battle was to signify the triumph of good over evil.

VIEW THE MAHABHARATA EPIC HERE
Video: https://www.youtube.com/watch?v=Xx4H_yuZbmU

By now, you might be wondering what the Mahabharata epic has to do with Indian beliefs about the body and soul? "It's a nice story, but I don't really understand what we can learn from it!"

The Mahabharata might be an old, epic fable, but it essentially summarizes the events that led to the development of ancient Indian civilizations. We can, therefore, learn a great deal by studying this ancient epic, as well as other scriptures, such as the Vedas, to determine how the ancient Indian and modern Indian civilizations have adapted and changed throughout the centuries.

Scholars use the term Santana Dharma in preference to the more commonly used name, Hinduism. In simple words, it can be stated as the ever-present, eternal law, or duty of everything in the universe of God. It is an obligation of every living and nonliving structure to perform its assigned role to keep the universe in a state of harmony. For humans, it requires adherence to a set of moral principles. It believes that every substance in the universe has a specific role and duty assigned to it.

God, or Brahman, is considered the ultimate true reality. Everything originates from the Brahman. The soul is a part of this ultimate true reality. The purpose or goal in life is to understand the ultimate truth. The soul is neither born nor dies; it simply inhabits a body. The soul, when it achieves the highest level of understanding and performing its assigned obligations, becomes a part of the ultimate truth - the Brahman. Studying, understanding and practicing the path of service to the universe leads to liberation from worldly life to become a part of Brahman.

MARRIAGE AND INHERITED CHARACTERISTICS

Hindu scriptures describe in detail the process of choosing a spouse, specifically instructing that the person's family must be screened for any known inherited diseases, going back ten generations. The ancient texts also advise evaluating both the maternal and paternal sides for good reputation, solid character, and a record of worthy deeds.

The *Laws of Manu,* the code of conduct for Hindus, make a special point of mentioning that illnesses such as hemorrhoids, consumption, dyspepsia, epilepsy, and leprosy are to be avoided in the families of the prospective bride and groom.

They further state: "A man of base descent inherits the bad characters of his father or his mother or both; he can never escape his origin." Another passage reads: "A woman always gives birth to a son who is endowed with the characters of his father."

From these scriptures, we learn that ancient Indians (Hindus) already had a sense that some characteristics, traits and diseases could be transmitted from generation to generation. This knowledge has been preserved for thousands of years, and is still in use over much of India today, even among the most remote, uneducated village inhabitants. Hindu families take great care in selecting a future spouse for their child, by carefully investigating and analyzing the future bride/groom's family line. They are often careful not to select a future husband/wife for their daughter/son if genetic traits and illnesses are evidently present, for fear their grandchildren will inherit the same traits.

THE HINDUS' BELIEFS ON BIRTH

The Upanishads, another set of the Hindu scriptures, cover conception and birth. One of the oldest is *Altareya Upanishad*. It describes the formation of a baby:

1. *In the male first the unborn child becometh. This which is seed is the force and heat of him that from all parts of the creature draweth together for becoming; therefore he beareth himself in himself, and when he casteth it into the woman, 'tis himself he begetteth. And this is the first birth of the Spirit.*
2. *It becometh one self with the woman, therefore it doeth her no hurt and she cherisheth this self of her husband that hath got into her womb.*
3. *She the cherisher must be cherished. So the woman beareth the unborn child and the man cherisheth the boy even from the beginning ere it is born. And whereas he cherisheth the boy ere it is born, 'tis verily himself that he cherisheth for the continuance of these worlds and these peoples; for 'tis even thus the thread of these worlds spinneth on unbroken. And this is the second birth of the Spirit.*

The Upanishad views on conception or pregnancy:

>Garbha Upanishad
>*It has white, red, black, smoky gray, yellow, tawny, and pale as the colors. What are the seven dhātus (tissues) when Devadatta (any person) desires enjoyment of objects? From the proper combination of qualities, six types of taste (rasa) emerge. From relish of food, blood is created, from it flesh, thence fat, bones, marrow, semen. By the combination of semen and blood the embryo (garbha) is born, and the heart regulates its growth.*

Charaka, who lived in the third century B.C, is regarded as the father of the traditional Indian system of medicine known as Ayurveda. His compendium on medicine, the Charaka samahita, states that the features of a child are derived from the reproductive materials of both parents, the food consumed by the pregnant mother and the soul, which enters the body.

THE HINDUS' BELIEFS ON LIFE AFTER DEATH

The general belief of Hinduism is of an eternal soul, which leaves the physical body when it dies and is reborn as a different being, in a cycle called "samsara". This cycle of rebirth of the soul continues till the body which carries it becomes pure, by learning to lead a life according to the teachings of the scriptures. It then attains salvation or "moksha" and becomes a part of "Brahman" or God and is liberated from the cycle of rebirth.

IN CONCLUSION

What we have learned of Hindu mythology and the ancient writings so far are that ancient Indian literature suggests that energy, physical strength and mental characteristics can be inherited. Ancient texts indicate that they had a wealth of knowledge about the traits that could be handed down from one generation to another. This knowledge is still present in modern Hindu society, as evidenced through the amount of time parents devote to finding a suitable match for their beloved children. A partner is carefully selected after it has been established that his/her family line has no unwanted traits that could be handed down to the next generation, and that the future husband and wife would be compatible as a couple.

SOURCES & REFERENCES – CHAPTER 3

Tyagi, Ashok. *All About Hinduism (From Vedas to Devas and Past to Present).* Shipra Publications New Delhi India 2015

CHAPTER 4:
ANCIENT JUDEO-CHRISTIAN BELIEFS ABOUT BIRTH AND THE HUMAN BODY

"Then Lord God formed man of dust from the ground, and breathed into his nostrils the breath of life; and man became a living being."

- Genesis 2:7

Our knowledge we shared thus far of the ancient Egyptians and Hindus has already provided evidence that knowledge regarding the origins of the human race and its procreation abilities was more common than might be supposed. But was the ancient knowledge of the human body, and the inheritance of parents' traits lost after the great civilizations, such as Egypt, came to an end? Or was the knowledge of the ancients known to other nations, such as the Hebrews?
Let's explore:

ANCIENT JUDEO-CHRISTIAN BELIEFS

The subtle links and shared history among us as human beings (no matter our race, religion or beliefs), are far more interlinked than most modern individuals think. The Old Testament is known today to most Christians, as the Hebrew Bible and Talmud, containing a collection of rabbinical teachings which form the basis of the beliefs and practices of Judaism (from which Christianity originated, by the way).

JUDEO-CHRISTIAN BELIEFS ABOUT THE HUMAN SOUL

In ancient times, people known as the Hebrews believed that humans were created, *'in the image and likeness of God'*. (Genesis 1:26-27). *'Our life-breath is the breath of God Himself* (Genesis 2:7).' We learn as much from The Talmud and the Bible. This divine 'inspiration,' identified as either the soul or the spirit of God, is said to have created human life from its earliest beginnings, which is what we would call fertilization or conception. Whenever we study the books of the Bible or Talmud very closely, we witness an astonishing array of evidence showcasing the ancient Hebrews ('Judeo-Christians') knowledge of heredity and inheritance. The Psalms, for example, read: *'thy hands have made and fashioned me,"* (Ps 118/119:73), and, *'thou didst form my inward parts; thou didst knit me together in my mother's womb.* (Ps 138/139:13).

In another passage, Job reminds God that his hands have fashioned him "from clay" and "knit [him] together with bones and sinews". *'God creates the "heart" of all persons and animates them with His own life-breath.'* (Isa 33/34:15; Ps 32:15, etc.). *'Man and woman are created on the sixth day.'* (Genesis 1:27 & 2:7)

And God said:
> "Let us make man in our image, after our likeness: and let them have dominion over the fish of the sea, and over the fowl of the air, and over the cattle, and over all the earth, and over every creeping thing that creepeth upon the earth. So God created man in his own image, in the image of God created he him; male and female created he them." (Genesis 1:26-27)
>
> And the LORD God formed man of the dust of the ground, and breathed into his nostrils the breath of life; and man became a living soul. And the LORD God planted a garden eastward in Eden; and there he put the man whom he had formed. (Genesis 2:7-8)
>
> And the LORD God said: "It is not good that the man should be alone; I will make him a help for him." (Genesis 2:18)
>
> And the LORD God caused a deep sleep to fall upon Adam, and he slept: and he took one of his ribs, and closed up the flesh instead thereof; And the rib, which the LORD God had taken from man, made he a woman, and brought her unto the man.(Genesis 2:21-22)
>
> This is the book of the generations of Adam. In the day that God created man, in the likeness of God made he him; Male and female created he them; and blessed them, and called their name Adam, in the day when they were created. (Genesis 5:1-2)

Some of the concepts, principles and observations in the Talmud and the Bible reveal that our ancestors had considerable awareness of the natural world that existed around them. In the Talmud, we read in Yevamot 64b about the circumcision of a newborn baby boy. The passage indicates that it is forbidden to perform the procedure if two sisters of the boy's mother had previously lost baby boys because of bleeding when circumcision was performed. The same exemption is not extended to the father's side of the family. This clearly reveals that they had an idea that a bleeding disorder (now recognized as hemophilia) was passed on from generation to generation from the mother's side of the family and not the fathers. This was an enigmatic puzzle, only solved by modern science after the discovery of X and Y chromosomes, with the gene for hemophilia being located on the X chromosome and, hence, passed on from the mother's side to the male child.

Genesis, Chapter 30 describes a transaction that took place between Jacob and his father-in-law, Laban. Jacob leaves his family home to stay with Laban in search of a wife and ends up caring for Laban's flock of sheep. In exchange for taking care of his father-in-law's sheep, Jacob agrees to accept all the speckled and spotted animals; while those with a uniform color are to go to Laban. But, as with many incidents in life, Laban becomes dishonest and cheats. However, Jacob is not defeated and cleverly out-maneuvers Laban by applying some rather interesting genetic 'manipulation' tactics to ensure he gets his fair share. In the process, Jacob's story is recorded in The Talmud, and becomes one of the first actual written proofs of genetic knowledge in the ancient Hebrew world.

JACOB'S STORY – HIS KNOWLEDGE OF GENETICS

Jacob Prospers

₂₅ Now after Rachel had given birth to Joseph, Jacob said to Laban, "Send me on my way so I can return to my homeland. ₂₆ Give me my wives and children, for whom I have served you, and I will go on my way. You know how hard I have worked for you."

₂₇ But Laban replied, "If I have found favor in your eyes, please stay. I have learned by divination that the LORD has blessed me because of you." ₂₈ And he added, "Name your wages, and I will pay them."

₂₉Then Jacob answered, "You know how I have served you and how well your livestock has fared under my care. ₃₀ Indeed, you had very little before my arrival, but now your wealth has increased many times over. The LORD has blessed you wherever I set foot. But now, when may I also provide for my own household?"

₃₁ *"What can I give you?" Laban asked.*

"You do not need to give me anything," Jacob replied. "If you do this one thing for me, I will keep on shepherding and keeping your flock. ₃₂ Let me go through all your flocks today and remove from them every speckled or spotted sheep, every dark-colored lamb, and every spotted or speckled goat. These will be my wages. ₃₃ So my honesty will testify for me when you come to check on my wages in the future. If I have any goats that are not speckled or spotted, or any lambs that are not dark-colored, they will be considered stolen."

₃₄ *"Agreed," said Laban. "Let it be as you have said."*

₃₅ That very day Laban removed all the streaked or spotted male goats and every speckled or spotted female goat—every one that had any white on it—and every dark-colored lamb, and he placed them under the care of his sons. ₃₆ Then he put a three-day journey between himself and Jacob, while Jacob was shepherding the rest of Laban's flocks.

₃₇ Jacob, however, took fresh branches of poplar, almond, and plane trees, and peeled the bark, exposing the white inner wood of the branches. ₃₈ Then he set the peeled branches in the watering troughs in front of the flocks coming in to drink. So, when the flocks were in heat and came to drink, ₃₉ they mated in front of the branches. And they bore young that were streaked or speckled or spotted. ₄₀ Jacob set apart the young but made the rest face the streaked dark-colored sheep in Laban's flocks.

Then he set his own stock apart and did not put them with Laban's animals.

41 Whenever the stronger females of the flock were in heat, Jacob would place the branches in the troughs, in full view of the animals, so they would breed in front of the branches. 42 But if the animals were weak, he did not set out the branches. So the weaker animals went to Laban and the stronger ones to Jacob.

43 Thus Jacob became exceedingly prosperous. He owned large flocks, maidservants and menservants, and camels and donkeys.

SOURCE: The Berean Bible (www.Berean.Bible) Berean Study Bible (BSB) © 2016, 2018 by Bible Hub and Berean.Bible. Used by Permission. All rights Reserved. Free downloads and licensing available.

What a phenomenal recorded story, isn't it? And one of the very first recorded accounts we have of 'genetics and heredity' in the ancient world. More amazingly, this story dates back to as early as the second millennium BC and is thought to originate from North West Mesopotamia.

DID JACOB USE GENETICS TO CREATE THE TYPE OF GOATS & SHEEP HE WANTED?

In many ways, Jacob's 'breeding program' might seem a mere fairytale; yet, is there evidence as to why his plan worked?

MD's, Alex C. Perdomo and Georgina Chan Perdomo have investigated the story of Jacob in greater detail, in search of clues as to why and how Jacob managed to create spotted sheep and goats after his father-in-law, Laban, removed the spotted rams so that these could not mate with the plain white or brown sheep and goats he left in Jacob's care. Laban thought that he understood the natural laws of inheritance, that spotted sheep would only breed spotted ones and uniform colored would breed uniform colored ones. However, Jacob knew better that inheritance was more complex, with recessive and dominant genes and mating can produce different combinations.

Interestingly enough, Drs. Alex and Georgina found upon further investigation that there might have been a reason why Jacob used only fresh poplar, almond and plane tree branches, and specifically placed them in the drinking troughs of the animals.

As we all know, some foods stimulate certain aspects within us as humans. Garlic is reported to help blood flow, lower cholesterol and keep the heart healthy. Salmon and oysters may help increase libido. Castor oil is considered to have anti-inflammatory and antibacterial properties that help heal skin, promote hair growth, treat infections and ringworms, and can be used as a laxative to remove toxins from the body.

Almond on the other hand, is said to improve fertility, and strangely enough, the simple aroma derived is considered to stimulate arousal in women, increasing their libido. Perhaps this is the reason Jacob made use of almond branches?

ANCIENT JUDEO-CHRISTIAN BELIEFS ABOUT BIRTH

Followers in the ancient world of what became the Judeo-Christian tradition regarded birth as a serious matter. In scripture, we read: "YHWH (Yahweh, the god of the Israelites) gasps and pants like a woman in childbirth" (Isa 42:14), while "pain" and "anguish" are similar for men hearing the news of a war and women in labor, according to Jeremiah. (Jer. 6:24, Jer. 50:43).

In The Hebrew Bible, we find an interesting play of words as the scriptures describe conception as *the sowing of seeds*. When talking about the embryo, there is a comparison that the womb has the depth and darkness of Earth. The fetus was thought to be like woolen threads that were later knitted by a divine force (Ps 139:13, Job 10:11).

When it came to the creation of life, the ancient Hebrew writings describe the process as similar to the way a potter forms clay. (Job 10:19). Firstly, the clay is mixed and molded together, then formed into a ball before being worked outwardly into a desired product. The process is similar to the way a man's sperm is 'mixed' together with the woman's egg to later form the fetus.

A variety of birth rituals were performed once a child was born into a Hebrew family. One such is the cutting of the umbilical cord. Thereafter, the child is often rubbed with salt. Although we are not sure why specifically salt was used, it might be due to the disinfectant properties of the substance, as well as the esteem with which salt was regarded in those times; salt was an incredibly valuable commodity, often used in transactions.

Both parents would name the child during a ceremony, with baby boys being circumcised, preferably by the father, on the eighth day. (Lev 12).

ANCIENT JUDEO-CHRISTIAN BELIEFS ON LIFE AFTER DEATH

"The dead do not praise the Lord, nor do any that go down into silence." (Psalm 115)

The ancient Greeks and Hebrews believed that the dead and living existed in a communal / common space. The dead were said to have a shadowy fourth-dimensional existence. This was known as a parallel dimension. The dimensions were thought to exist at the same time, and one would pass from the living dimension to the dimension of the dead upon death, when the soul left the body.

In The Hebrew Bible, the place of the dead had an actual name; '*Sheol*,' (similar to Hades in the Greek mythology). It's important to note that Sheol was not only the dimension where the sinners would go, but was considered the ultimate fate of all humanity, no matter the nature of their deeds.

As miserable as it was, the afterlife was an accepted doctrine in the lives of the ancient Hebrews, with most people believing that, no matter what they did in their life on Earth, when they died they would go to Sheol. Luckily, the perspective changed a few centuries before the birth of Jesus Christ. Sheol became a place where the 'bad' or wicked would go, while the 'good people' had a chance to enter 'Heaven'. (Wisdom and Maccabees provide more information regarding the morality changes that occurred, and how the ancient Hebrews now started to view life by distinguishing 'good acts' from 'bad acts'.)

So, what exactly happened to a Hebrew upon death, after Sheol no longer became the only option to mortals? Christian teachings indicate the existence of differing places for the righteous and the wicked. An individual would be brought to whichever space they belonged to after the last judgment, as mentioned in Matt 16:17-20. In Luke 16, it's interesting to note that being 'rich' was also regarded as a sin, in that rich men would be punished, while the poor would be rewarded by everlasting life in heaven. A detailed story account tells about a poor man known as Lazarus who meets Abraham and finally settles with him as his reward. But, as with many controversies in scriptures in any religion, there were 'exceptions' to the rule, as we later learn even rich men could go to heaven if they showcased good acts during a lifetime on Earth.

WHAT HAPPENED TO THE 'WICKED'?

After the concept of Sheol changed, the ancient followers of the Judeo-Christian tradition believed in the afterlife, in either a good form or eternal torment. They believed in the final judgment. (Yet, it doesn't happen immediately after death; It's a future event.) (Matt 8:12, Matt 13:42, Matt 24:51, Rev 19:19-21, and Rev 20:7-15) These ideas of the afterlife later developed into the concepts of heaven and hell, to ensure the good people are rewarded while the 'bad' are punished. Early Christians harmonized these concepts, as evident from the New Testament Gospels.

IN CONCLUSION

Ancient Judeo-Christian writings tell an in-depth story about the process of life and death. Verse after verse explains in explicit detail the different stages of life in the womb of a woman, from the first step as to how the embryo develops, to the weaving of the fetus by a divine knitter into its final human form.

Various biblical figures, such as Jacob and Laban, show that they had knowledge about genetics and inheritance and used this knowledge in their everyday lives.

Studying these divinely inspired texts clearly suggests that ancient Judeo-Christians understood the world around them, even though they didn't bother to investigate the phenomena governing them. Instead, they chose to *believe* and have faith in divine intervention.

SOURCES & REFERENCES – CHAPTER 4

Perdomo, A.C. (MD), & Perdomo G.C. (MD). *Is DNA in the bible?* ETZ Yoseph. [Online] Available at: http://etzyoseph.org/is-dna-in-the-bible/

CHAPTER 5:
ISLAM'S BELIEFS ABOUT BIRTH AND THE HUMAN BODY

"Verily, we created man from a drop of mingled fluid" - Al Quran 76:2 (Translation: Yusuf Ali)

Not only did the ancient Egyptians, Hindus and Hebrews have knowledge of embryology, human body and inheritance, but we come to know upon further investigation that Islam had just as much insight, if not more, into the matter as other ancient civilizations.
Let's have a look:

FORMATION OF A BABY

"We created humans from a drop of mingled fluids," Allah says in The Holy Quran. This verse clearly indicates that both the male and female participate in the formation of the embryo.

But The Holy Quran is not the only formidable written resource indicating knowledge on heredity; The Hadith, the teachings of holy Prophet Muhammad(PBUH)[1], also explain inheritance of specific traits. For example, a man once asked the Prophet why his wife had given birth to a 'black' child. Prophet Muhammad (PBUH) proceeds by first asking some qualifying questions:

"Do you have any camels?"
"Yes," Replied the man.
"What is their color?"
"Red," said the man.
"Are there any dusky ones among them?"
"Yes," answered the man. "But how is this connected to my child being born black?"
"It is perhaps the strain to which the child has reverted back," answered Prophet Muhammad (PBUH).

Wow! Isn't that a profound selection of words the prophet chose to explain to us the reason behind the child's color? Through reading this story, we come to understand that The Hadith clearly talks about genetic traits that may be suppressed in one generation and appear in another. Yet, should we still be surprised about the knowledge of the ancients after all we have learned from ancient text thus far?

As with all things, humans often forget about history, and soon 'all is forgotten,' as is said very wisely in the opening passages of one of the most well-

---

[1] Peace Be Upon Him (PBUH)

constructed and entertaining books and movies in history, *The Lord of the Rings – The Fellowship of the Ring*:

"Much that once was, is lost.
For none now live who remember." (J.R.R Tolkien)

Luckily, we have ancient texts and accounts of history to remind us about the knowledge of our ancient ancestors so that we never forget. As such, I came to learn through various studies that the ancients had achieved major feats we have forgotten today. One such was the major advance made by the ancient Muslims.

Islam was once the intellectual center of the world for science, medicine, astronomy, mathematics, philosophy, and chemistry (Syed, 2018). It was a time when Muslims had established one of the largest empires in history. Some of the greatest minds came from the Islamic world, all of which carefully analyzed the natural world around them as objectively as they could. This is interesting to note, as we later learn that this scientific method of enquiry for some reason later ceased to exist, as religion, myth and mythology started taking over major parts of the world.

The Muslim ruler, also known as Caliph, Harun al-Rashid in the eighth century established a " House of Wisdom" in the capital city, Baghdad. Baghdad was the crossroads of trade and travel routes from Europe to Asia. The house of wisdom became a center of activity, attracting Muslim and Non-Muslim scholars. There was a burst of intellectual activity from translation of ancient texts, synthesizing information from various sources to innovation and new discoveries. It created a unique culture of bringing together scholars of diverse backgrounds from China, India, the Middle East, Africa and Europe, creating a true Universal civilization. At the dawn of the Renaissance, the Europeans picked up the preserved and resurrected ancient knowledge and spirit of exploration started by the Muslims, to usher in the scientific revolution which continues to this day.

ISLAMIC BELIEFS ABOUT THE HUMAN SOUL

Islam preaches that the human 'soul' is an immaterial and unseen aspect of a human being. The soul is regarded as the body's essence and an integral part of life. Muslims believe that souls will be held after death and rejoined with their bodies on the *Day of Judgment*. (INTERESTING TO NOTE: This belief is very similar to the belief held by Christians that their body will be awakened on the *Day of Judgment* by God, and that their spirit will then be rejoined with their body to receive judgment for the actions they had done during their life on Earth.)

In Chapter 4 (Al-Nisa), verse 1 of the Quran, Allah reveals the true nature of the human soul: "O mankind reverence your Lord, who created you from a single soul and from it created its mate and from them hath scattered countless men and women." In Surah Al-Hijr, verse 29, Allah tells the believers the story of creation and talks about the human soul in these words: "When I have fashioned him and breathed into him of My spirit, fall ye down in obeisance unto him." (It's interesting to note here the similarities that exist in The Bible / Talmud.)

According to the Quran, a soul can benefit from what a human does in this world. So, it's possible to have a 'good' or 'evil' soul, based on the kind of energy it possesses. In chapter 41, verse 46, Allah explains this very nature of the soul to believers: *"Whoever works righteousness benefits his own soul, whoever works evil; it is against his own soul."*

Although Islam does talk about the soul at times, it also suggests that the understanding of the soul is something that's beyond the human mind. It's something that the human mind is technically incapable of understanding fully. The knowledge of the soul is restricted, as it's a divine matter. In chapter 17, verse 85, Allah says in the Quran: *"And they ask you (Prophet Muhammad (PBUH)) about the soul. Say: The soul is one of the commands of my Lord, and you are not given aught of knowledge but a little."*

ISLAMIC BELIEFS ON BIRTH

The teachings indicate the belief that life does not necessarily begin at the time of conception. It begins when Allah places the soul (Ruh) in the human body. However, life as a whole didn't begin with the creation of humans. In Chapter 76, verse 1, Allah says in the Quran: *"There was a time when man was non-existent and he was not even mentioned."* Similarly, in Chapter 4, verse 1, the Quran claims that God created a single soul and then its mate. From there, humans spread over the planet as men and women.

Allah explains it again in Chapter 7, verse 189 in these words: "It is He who created you from a single person, and made his mate of like nature, in order that he might dwell with her. When they are united, she bears a light burden and carries it about unnoticed. When she grows heavy, they both pray to Allah their Lord…"

This verse not only explains that all souls were created from a single soul but also talks about human life inside the womb. Birth is explained in various settings. In one hadith, Prophet Muhammad (PBUH) explains:

> *"The creation of each one of you is brought in the womb of his mother for forty days as a **germ cell**. Then for a similar period, he is in an **embryonic lump**, hanging like a **leech**. For another forty days, he is a **mudghah** (something chewed), then an angel is sent and he blows in him **the ruh** (soul)."*

This clearly explains different embryotic stages of human life – something that science has only discovered recently.

ISLAMIC BELIEFS ON LIFE AFTER DEATH

In Chapter 10, verse 56, God (Allah) talks about life, death, and the Day of Judgment as follows: "He gives life and causes death. And to Him, you will be returned."

There is a belief that life exists after death, as well as a *Day of Judgment*. It is the 'day of reckoning', where every soul will be asked to answer for the deeds they committed during their life on Earth; the good, the bad and the ugly (Islam-guide.com, 2018). In Chapter 6, verse 164 in the Quran, Allah tells humans about their eventual fate: *"Every soul earneth only on its own account, no bearer of burdens can bear the burden of another. Your goal in the end is towards God: He will tell you the truth of things therein ye disputed."*

The Quran further discusses the state of affairs after death in these words in Chapter 3, verse 30:

"On the day when every soul will be confronted with all the good it has done, and all the evil it has done, it will wish there were a great distance between it and its evil."

The virtuous will be rewarded with eternal happiness in heaven, while the evil doers will not escape Allah's long arm of justice, condemned to severe punishment in hell-

"God created the heavens and the earth for a true purpose: to reward each soul according to its deeds. They will not be wronged" (45:22)

Early Muslim physicians further advanced understanding of inheritance, like Abu al-Zahrawi, known to the west as Albucasis, who is considered to be the first to clearly detail the hereditary nature of the bleeding disorder, hemophilia.

IN CONCLUSION

Islamic views suggest that all human souls originated from a single source. Today, we can trace this by studying human DNA. Further important to note, is the Islamic belief in the *Day of Judgment*. As a result, all humans will be judged for their deeds on the *Day of Resurrection*.

Although Islamic views in modern society have been clouded by various oversights and unfortunate extravagant dismissals of modern science, we come to discover through the study of the words in the Quran and of Prophet Mohammed (Peace be upon him) that he had in-depth insight about embryology, the interior of the womb of a woman and what the fetus looked like, as well as the fact that traits that can be inherited even some generations later. All of this knowledge was known during the 6th century; that is, almost 1,500 years before the development of ultra-sound, or before modern scientists concluded that DNA is responsible for traits such as albinism, that could be reflected back several generations later.

SOURCES & REFERENCES – CHAPTER 5

Syed, I. (2018). *The Nature Of Soul: Islamic And Scientific Views.* [online] Irfi.org. Available at: http://www.irfi.org/articles/articles_51_100/nature_of_soul.htm [Accessed 22 May 2018].

Islam-guide.com. (2018). *Islam Guide: Life After Death.* [online] Available at: https://www.islam-guide.com/life-after-death-by-wamy.htm [Accessed 22 May 2018].

Ghareeb, Bilal AA. "Human genetics and Islam: scientific and medical aspects." *The Journal of IMA* 43, no. 2 (2011): 83.

CHAPTER 6:
ANCIENT CHINESE BELIEFS ABOUT BIRTH AND THE HUMAN BODY

"The idea of creation out of nothing has ever remained entirely foreign to the Chinese mind, so much so that there is no word in the language to express the idea of creation."
- E.J. Eitel

From the shocking findings that we had 'geneticist knowledge and expertise' in existence since the earliest recorded times of the human race, we go on to further explore the knowledge of the ancient Chinese and their views on the human body and inheritance of character.

Let's explore:

THE ANCIENT CHINESE

"Dragon gives birth to dragon, phoenix gives birth to phoenix, and a thief has a son who is a burglar," is an old Chinese saying.

The ancient Chinese felt that traits that are inherited and those that are not inherited couldn't be distinguished by daily human experience. In their history spanning over 5000 years, the ancient Chinese developed a unique culture based on the teachings of Confucius, Tao, and Buddha. They believed (and believe to this day) in the concept of a vital energy known as *qi*. The understanding is that everything in the universe is simply a manifestation of various forms of *qi*.

Qi comes in two opposite forms – *yin* and *yang*. But, there is an interaction and exchange that occurs between yin and yang in the universe. According to the Chinese belief, everything as we know it is a result of this interaction between the two opposites. From as early as the 3rd century BC, the ancient Chinese believed in the workings of *yin* and *yang*. They then used their observations to explain some of the observed facts of heredity and inheritance in their immediate societies. Ancient Chinese faith indicates a belief system based on the gods, spirits and ancestors, which can all impact life on Earth. These elements could easily affect the outcome of both human effort and divine order, such as childbirth, crops, and warfare. It's also important to note that some ancient Chinese cultures were matrilineal, meaning women were in control of the familial line, and probably had different principles for inheritance compared with other cultures that were patrilineal or male-dominated.

Interestingly enough, the ancient Chinese noted that each species would only give birth to more of the same and not any other type of creature. For example, humans didn't give birth to dragons, and dragons didn't give birth to dogs. They instead observed that dragons gave birth to dragons, while the son of a thief gave birth to a burglar. While the comparison of a dragon giving birth to a dragon is pretty straight forward (the genetic code of one species being carried over to the next), why the sudden observation that the son of a thief would give birth to a burglar?

This is a very important observation, especially since it is regarded that modern science is responsible for the discovery of how genetic traits are indeed transmitted from generation to generation through the 'code' locked in our DNA. Yet, how did the ancient Chinese know that these traits, like criminal behavior, among many others such as medical problems, were handed down from one generation to the next?

ANCIENT CHINESE BELIEFS ABOUT THE HUMAN SOUL

Ghost stories are abundant in the ancient Chinese belief system. Even today, the Chinese have various tales of ghosts and festivals to commemorate these stories.

According to ancient Chinese belief, ghosts (traditionally *guei* and *kuei*) are spirits of deceased people who didn't receive a proper burial. These spirits didn't want to leave the dimension of planet Earth and played a very important role in Chinese religion and culture. It is said that the spirit of a deceased person could reveal the future. Some selected individuals are able to contact the spirits and ask them questions about the future. Spirits had a special place in ancient Chinese culture, as they believed that ancestral spirits were connected with gods and other eternal beings. The ancient Chinese further believed that everything in the universe was a result of two spirits interacting with each other. These spirits, known as *Yin* and *Yang*, were considered responsible for all of creation - including the human soul.

ANCIENT CHINESE BELIEFS ON BIRTH

Nuwa, the goddess of humankind, molded humans from the mud of the Yellow River. She created humans and gave them life until she one day grew tired of the whole process. (Most probably the human race was populating too fast for her to keep up.) Nuwa then decided it was time humanity took on some more responsibility, so she gave humans the power to reproduce without requiring her to make and breathe life into them (Asia Society, 2018).

A WOMAN'S INFLUENCE ON AN UNBORN CHILD

Chinese beliefs indicate that a pregnant woman could influence a child in her womb through her thoughts. A very old tradition explains this concept as follows: *"What affects a woman's mind will also affect her heart and connect with the baby in her womb."*

The above is a very astute observation, especially since modern science has now established the same, and pregnant women today are advised to be considerate of how they interact and treat themselves while pregnant, as negative, adverse feelings and thoughts could influence the embryo in the womb.

ANCIENT CHINESE BELIEFS ON LIFE AFTER DEATH

Eternal life after death is a concept and belief held among most ancient Chinese. There was the belief that souls would move after death to another world, where they had to live a normal life, very much the same as the one they lived on Earth. There were everyday problems in this new world, with some even having to earn money. Even today, you will notice some Chinese burn paper and other items to give it to their ancestors who may need these things in another realm (National Library Board, 2018).

Another ancient Chinese story on life after death, states that souls remain in the body after death. After some time, the soul travels on the road to rebirth and meets an old goddess, who will serve a forgetting soup to the souls so that they can leave behind their memories when they move to another realm. But, all the soul's deeds are recorded on a magic stone before they are reborn.

After Buddhism came to China via what was then famously known as *The Silk Route*, the concept of the realm of the dead became somewhat refined with various theories of reincarnation, judgment, rewards, and punishment incorporated. According to ancient Chinese Buddhism, every person is transferred to *Yama-Rajas* (Kings of Hell) after death. It's here an individual answers for the deeds committed during its lifetime on Earth – the good and bad. *Yama-Rajas* listens carefully and then judges the person according to his/her deeds.

It is important to note here that Taoists believe in immortality that comes after this life. However, reaching an immortal stage after death isn't for everyone. One has to perform various tasks and rituals and live life in a particular way. Only then will he or she be blessed with immortality.

IN CONCLUSION

Ancient Chinese religion talked in detail about soul, birth, and life after death. They worshipped many gods and spirits, where the dragon spirit was the most popular. Their belief in immortality and a judgment day has shaped what is known as modern Chinese culture today.

The rich Chinese culture is also heavily influenced by famous philosophers such as Confucius, Laozi (Taoism), and Buddha. These beliefs helped the Chinese shape their history and culture to become one of the world leaders in technology and science in modern times.

SOURCES & REFERENCES – CHAPTER 6

National Library Board, Singapore. (2018). *Chinese birth rituals* | Infopedia. [online] Eresources.nlb.gov.sg. Available at:
http://eresources.nlb.gov.sg/infopedia/articles/SIP_2013-05-14_113920.html
[Accessed 24 May 2018].

Thich Thanh Nguyen 2018. *Death and Rebirth.* Tuvienquangduc.com.au. [online] Available at:
http://www.tuvienquangduc.com.au/English/rebirth/28afterlifechina.html
[Accessed 24 May 2018].

Encyclopedia.com. (2018). *Afterlife: Chinese Concepts* | Encyclopedia.com: FREE online dictionary. [online] Available at: https://www.encyclopedia.com/environment/encyclopedias-almanacs-transcripts-and-maps/afterlife-chinese-concepts [Accessed 24 May 2018].

Deathreference.com. (2018). *Chinese Beliefs - rituals, world, burial, body, funeral, life, customs, history, time, person*. [online] Available at: http://www.deathreference.com/Ce-Da/Chinese-Beliefs.html [Accessed 24 May 2018].

Asia Society. (2018). *Chinese Religions and Philosophies*. [online] Available at: https://asiasociety.org/chinese-religions-and-philosophies [Accessed 24 May 2018].

CHAPTER 7:
LIKE FATHER, LIKE SON - ARISTOTLE, THE GREEKS, AND THE DAWN OF RATIONAL THINKING

"Reason is immortal, all else is mortal."

– Pythagoras

LOGOS VS. MYTHOS

Many ancient civilizations intuitively understood the continuity of life and some of the mechanism for passing along physical, emotional, and spiritual traits. The majority of the populace, however, ultimately accepted these as essentially divine providences. But, with the rise of Greek civilization, a new way of thinking was introduced. In the sixth to fifth century BC, Greek philosophers and physicians began to ask the first concrete questions concerning the laws that govern living creatures, their development, and creation. Today, many believe that this scientific approach to nature was perhaps the greatest breakthrough ever in human thought. For the first time, the best and the brightest minds were asking questions about unifying themes that could explain the behavior of nature without reference to religion, myth or mythology; questions such as: *"How is life created? Why do dogs only give birth to dogs and not dragons? Why is the nose of her grandmother present on her face two generations later?"*

WHAT HAPPENED IN GREECE AND HOW IT BECAME THE CENTER OF WISDOM AND KNOWLEDGE

Greece is a small country with many islands and several peninsulas that influenced the development of city-states. The first settlers arrived in the northern part about 6,500 BC. It is believed that the Greek language was brought by an Aryan race of people, invading ancient Greece from the Russian steppes around 2500 BC. As a civilization, the Greeks were responsible for a hotlist of new discoveries that would ultimately influence the whole world.

Let's, therefore, investigate the top three most important influences left behind by the ancient Greeks:

ARISTOCRATIC RULE WAS REPLACED FOR THE FIRST TIME

Universal aristocratic rule, with its class distinction, was the norm in the ancient world, but the system came to an end when it was replaced for the first time by a democratic form of government. This marked the beginning of the *Golden Age of the Greeks* in the sixth century B.C. This important political step is what would lead, in due course, to humans investigating how religion, myth, mythology and politics fit into the nature of life, and if it had any place at all in the way life was formed. Lower classes were, for the first time, given the opportunity to gain education and knowledge; something that, in Greece, had previously been reserved for only the upper classes and royalty.

THE GREEKS RELIED ON OBSERVATION AND ANALOGICAL DEDUCTIONS

The ancient Greeks were the first to truly understand and propose that the world may be known through careful observation and logical deductions. They laid the foundation to systematic study and careful observation without the aid of a god and divine intervention.

Philosophical speculation eventually led to the formulation of scientific hypotheses regarding reproduction and heredity. This is an important step forward, since most of the ancient world was, at that time, ruled by myth and mythology when it came to birth, procreation and heredity.

SOCRATES AND PLATO – THE BEGINNING OF 'REASON' IN GREECE

Socrates, considered by many to be the founding father of Western philosophy, was born in Greece in 469 BC. According to Socrates, *'ultimate wisdom came from knowing oneself.'* He reasoned that the more we knew about ourselves, the better we would be able to reason logically and objectively, and find the path to true happiness.

Plato ably carried forward the torch lit by his teacher, Socrates. He founded *The Academy* in Athens, the first institution of learning in the Western world. It taught astronomy, biology, mathematics, political theory and philosophy. Plato was an able teacher, but he greatly admired Socrates, who enjoyed conveying knowledge to his students. One of the best stories showcasing Socrates' ambition to educate all can be found in one of Plato's writings, *Meno*. Here, he describes a dialogue between Socrates and a slave boy.

One day, Socrates drew a square in the ground and asked the slave boy to double the area of the square. The slave boy answered: *"It is double the length of the sides."* Socrates pointed out that it would make the square four times larger. The slave boy tried again, guessing other lengths, but finally said: *"I give up."* Socrates then guided him by posing simple step-by-step questions to get to the correct answer. Upon enquiry by other students and scholars as to why he took the time to educate the slave boy in such a manner, Socrates explained that all learning is, in fact, recollection. We have innate knowledge; it simply needs to be recalled. A teacher can, therefore, guide a student in the process. (The Socratic method of teaching is especially popular in law schools today, where students are encouraged to develop critical thinking and engage in analytical discussions.)

Although Socrates is well known in modern times, perhaps the most important teaching he left behind is his philosophy that knowledge and science should focus on making society better. He attempted to establish a system based on reason, rather than theology.

As with all new sciences, it takes time for it to impact human society. Socrates' and Plato's teachings and findings were revolutionary for their time, often disregarded, ridiculed or misunderstood by their peers, and the majority of the populace continued to cling to myth and mythology to explain their existence.

Although many of their teachings were a step forward, it was only years later that society understood their importance. As happens today, a prophet, messiah or wise individual is often not respected or noted as worthy until well after their death. But we can surmise that life on earth improved from studying the work of those that have walked the path of objective, calculated reasoning long before us.

ARISTOTLE – INVESTIGATING THE GREATEST PHILOSOPHER OF ALL TIMES: HIS BELIEFS & THOUGHTS ABOUT GENETICS.

As if the findings and philosophies of Socrates and Plato were not already impressive and ahead of their time, a young man would take to the world stage in 367 BC, who would change, forever, the world as the ancients knew it,.

In the year 367 BC, from a small city in the region of Macedonia named Stagira, came a 17-year-old young man named Aristotle to enroll in Plato's *Academy* (Ancient Greek: Ἀκαδημία) (428/427 BC – 348/347 BC). Being the son of a physician by the name of Nicomachus - the physician to King Amyntas of Macedonia - Aristotle was exposed to the monarchy from an early age. However, he was not greatly inspired to continue in his father's footsteps, showing more interest in subjects such as mathematics, cosmology, logic, music and politics.

Today, Aristotle is known throughout the world for having shaped human thought for well over two thousand years. A prolific researcher and writer, he researched and wrote on a wide range of subjects, covering biology, mathematics, logic, cosmology, metaphysics, ethics, politics and music.

WHO INFLUENCED ARISTOTLE?

Before we review the contributions of Aristotle in more detail, it would be worthwhile to explore the Greek thinkers and philosophers who laid the foundation of transition from *Mythos* to *Logos* – Myth to Logic - in order for us to comprehend what type of knowledge and background education Aristotle received:

624 BC – 546 BC: THALES – EVERYTHING IS DERIVED FROM WATER

Thales was probably the oldest thinker in the history of Western civilization, from 624 BC through 546 BC. He was one of the seven sages of Greece, who traveled widely and brought the essentials of Geometry from Egypt. Thales proposed that the multiple phenomena of nature were derived from a single primal substance, water, which is capable of motion and change.

610 BC – 546 BC: ANAXIMANDER – AIR AS THE PRIMORDIAL ELEMENT

Anaximander, a pupil of Thales, detailed a theory of creation of the universe based on observation and rational thought. He proposed that living creatures and human beings evolved from water. Anaximander claimed that humans first developed inside the fish, from where they eventually were cast ashore. He wrote a treatise entitled, *On Nature*. He thought that the air, in which the heavenly bodies move about, is the primordial element. He speculated that all things were derived from air by the process of condensation and rarefaction.

535 BC – 475 BC: HERACLITUS – EVERYTHING IS CHANGING.

In the 6th century BC, Heraclitus, in his writing about nature, came up with the idea that everything in nature is constantly changing. This was a fundamental change in philosophical thought from " Being" to " Becoming". Nature is not as it is, but is in a constant state of change. He summarized this in his statement, which has become a famous quote "Ever-newer water flows on those who step into the same river". He also proposed the doctrine of opposites; it is from opposites, and not from unity, that harmony is produced.

He wrote that everything works on the basis of "Logos". The meaning of the word *logos* in ancient Greek has been interpreted in several ways, including 'principle', 'plan', 'measure', etc. He believed logos was always present, but humans did not understand and appreciate it; like a sleeping person, not sensing what is around.

492 BC – 432 BC: EMPEDOCLES – FORCES OF ATTRACTION AND REPULSION GIVE BIRTH TO LIFE

Empedocles (492-432 BC) proposed that four elements, earth, air, water and fire, made up everything. It was the first step in a universal theory of matter. He opined that, at one time, they were harmoniously intermingled in the universe, which was in the form of a sphere. It was the action of various forces of attraction and repulsion on the four elements that gradually produced the world, including animals and human beings.

460 BC – 370 BC: THE BEGINNING OF THE ATOMIST SCHOOL – EVERYTHING IS MADE OF 'ATOMS'

Zeno, Leucippus and the famous Democritus (460-370 BC), founded the atomist school. They described the world as being made up of an infinite number of extremely small qualitatively indistinguishable particles, called 'atoms.' According to these philosophers, these atoms arrange in different positions to create different living creatures.

342 BC – 271 BC: EPICURUS – THE CHAMPION OF GREEK MATERIALISM

Epicurus (342-271 BC) emphasized that there are eternal laws that govern the course of the world. Natural philosophy was to free man from the belief in supernatural forces, from the fear of death, the gods and from their concerns about an afterlife. Reason, he claimed, would lead humans to tranquility.

460 BC – 370 BC: HIPPOCRATES – THE FORMATION OF THE EMBRYO AND HEREDITY

The Greeks established a medical school in 600 BC. One of their most famous students, Hippocrates, created a medical oath which students all across the world subscribed to, even until today. The oath upholds ethical standards for the best interest of the patient.

Hippocrates was perhaps the first to observe the development of a chick embryo. He was also the first to write on the subject of heredity. In his *'Treatise on the Formation of the Embryo,'* he wrote that the sex of the offspring was a matter of chance. According to his thesis, the interaction of water and heat created the embryo, which initially is porous and homogenous but as it dries up from the heat, it solidifies.

Hippocrates believed that characteristics were inherited from parents. He taught that reproductive material, the semen, was received from all parts of the body; the testicles were merely reservoirs for the genetic material. Darwin later proposed the concept again, as the theory of pangenesis where each part of the body contributes to what Darwin called *gemmules*, which carry inheritable characteristics and are transmitted to the offspring during the reproductive process.

BACK TO ARISTOTLE AND HIS UNDERSTANDING OF INHERITANCE

Aristotle came into the world when Greek power and civilization were on the rise, and he took the pursuit of knowledge to its zenith. Aristotle was the first philosopher to separate various branches of philosophy, or natural sciences, as they are known today. He dissected animals, especially marine animals, and probably (it is speculated) dissected an aborted human embryo.

Aristotle started the system of classifying animals. He created two broad categories - animals that had blood and those that did not. For his immense contributions to the field, he is considered the father of biology.

Pythagoras had theorized that the male introduced moist vapors into the female during intercourse, which evolved into various body parts. Aristotle, not having the benefit of a microscope, believed that the sperm of the male had highly purified reproductive blood, containing nutrients from all parts of the body. He argued that the male's blood carried hereditary characters, which coagulated with the female's menstrual blood. The female's blood also had nutrients from various parts, but was weaker and hence the male contributed more to the offspring.

This concept of the interaction of maternal and paternal blood, though not accurate literally, has persisted as Aristotle's legacy in our language in the terms 'blood relatives', 'bloodlines' and 'one's own flesh and blood'.

Aristotle also came to the conclusion that inheritance was more a potential to inherit characteristics, rather than the absolute passing of characteristics from one generation to the next. Most of his ideas, far ahead of his time, are consistent with our current knowledge.

His work, mostly lost during the Roman era, was thankfully preserved by the Islamic caliphate of Baghdad and returned to Europe at the beginning of the Renaissance.

THE FORMATION OF AN EMBRYO

Throughout the ages, humans have been interested in exploring ways to select the gender of their offspring. At times, the desire to establish the gender of a child became such an obsession, especially among royal families desperate to produce a boy to one day rule in their stead, that some rather interesting notions and potions saw the light of day. Four thousand years ago, the Egyptians, for example, believed that women with greenish complexion mostly gave birth to boys. As can be expected, the said women were in great demand. Among Hebrews, the belief was to place the bed in a north-south direction; this practice would increase the chances of conceiving a male baby. (Leviticus – The Talmud.)

Things weren't much easier for aspiring parents in Greece. Aristotle believed that the male seed came from the right testicle and the female from the left. He was further of the opinion that the embryo was not pre-formed in the seed and simply grew to a larger size, but it developed as it grew, consistent with our current understanding. In his work, he describes the two models; preformation and epigenesis. Preformation theory involves the theory of pre-existence. In other words, preformation theory holds that a miniature individual already exists in the egg of the mother or the father's semen; when either one is stimulated, the child begins to grow.

Epigenesis, on the other hand, is of a different opinion, in that the embryo is an undifferentiated mass, and only starts expanding and gaining new parts during each development stage. Aristotle favored the theory of epigenesis. He also thought that the female contributed unorganized matter to the embryo, while the male contributed 'organized' matter, such as form and the soul. Furthermore, he was of the opinion the heart was the first element to form when a new embryo developed.

ROMAN RULE AND GREEK LEGACY

Civilizations come and go because of many factors, such as war, climate change, financial calamity or incompetent leadership. It was no different with the rise of ancient Greek knowledge and scientific study. When Alexander died, in 323 BC, it marked the beginning of the end of Greek power and dominance. Long and dreadful wars erupted between City States, with smaller kingdoms suffering the most as the economy declined. The power of the Romans rose and the unfortunate lack of unity amongst the Greeks allowed the Romans to overrun the Greek lands. At first, things stayed relatively unchanged under Roman rule, but later, Roman influences started to permeate the society in Greece - especially among science and academics. However, to this day, we can look back on the legacy left by Greek philosophers and scientists, such as Plato, Socrates and Aristotle. We know that their knowledge and ideas have been of great significance for the advancement of human civilization as a whole.

GALEN – THE EMPEROR OF MEDICINE

Under Roman rule, some elements of Greek culture, language and science continued to flourish. As such, one of the brightest minds to come forth after the period of Greek influence was Galen, born in 129 AD. In later years, he became the personal physician to Roman Emperor, Marcus Aurelius, known as *the Philosopher*, and the last of the five good emperors.

Luckily, we know a lot about Galen, due to 20,000 pages of his writing that survived. From his work, it is evident he was a prolific researcher and writer. Some of Galen's writings summarized previous Greek knowledge, while his medical textbook was the only comprehensive text on medicine until the *Canon of Medicine*, written by Avicenna, in 1025 AD. Galen performed surgical operations, dissected animals, and became known for 1,500 years as the *Emperor of Medicine,* due to his extensive medical knowledge and contributions to medical science.

When he wrote about reproduction, he didn't favor the religious belief that a divine creator was responsible for the formation of human beings. Instead, he tried to find the truth about reproduction by performing dissections on female apes. It was during one of these case studies that he noted a fluid in the uterus of a female ape. Galen called the fluid female semen or seed. In later years, he theorized that both males and females contributed equally to the physical attributes and character of their offspring.

IN CONCLUSION

What an exciting time it must have been! The state of knowledge that existed among Greek philosophers regarding heredity and genetics is astounding, especially since such knowledge existed well before the modern era, when science is said to have ultimately begun. Yet, what a wonderful world began to take shape as our ancient forebears began to systematically unravel the mysteries of birth and inheritance and to explore the very core of the innermost workings of living organisms.
Phenomenal!

SOURCES & REFERENCES – CHAPTER 7

Allen, Reginald E., ed. *The Greek Philosophers: From Thales to Aristotle.* The Free Press, New York, 1991.
Barnes, Jonathan, ed. *Complete works of Aristotle: The revised Oxford translation*. Princeton University Press, 1984.
Mattern, Susan P. *The prince of medicine: Galen in the Roman empire*. Oxford University Press, 2013.

CHAPTER 8:
THE RETURN OF MYTH & MYTHOLOGY; THE LONG DARK AGES

"Creeds made in the dark ages are like drawings made in the dark room."

- Joseph McCabe

Erroneously, we, as humans, often believe that the advances we make and the knowledge we gain will continue to grow and we will live better and more prosperous lives thereafter. Yet, the fact is that, just as we can gain knowledge, we can also lose it very easily.

Modern entrepreneur and inventor, Elon Musk, observed very accurately during an interview that it is a fact that humans can actually revert back to a previous state of mind, or lose the knowledge they gained. An excellent example is the way we managed to develop the technology to go beyond our planet and land on the moon in 1969. Yet, thereafter, it was as if the human mind stagnated and we didn't expand our space program to reach new and more exciting frontiers. The space program was actually put in the back seat for years, with NASA not gaining much ground to further the human's exploration of outer space until the revival of interest in space travel by Elon Musk in the last couple of years.

Ok, so what does this have to do with the long "Dark Ages" of the 5th to 15th centuries AD?

Actually, quite a lot. It proves the point that we, as humans, cannot always rely on the concept that evolution is something that continually moves towards a positive outcome. Evolution can actually be regressive as well.

As with many things in life, and as there is a season and a time for everything, the knowledge and wonderful advances made at the height of ancient Greek supremacy were lost during a long and treacherous period known as the long Dark Ages - lasting from the 5th to 15th centuries.

The Dark Ages were a period in which the pursuit of scientific knowledge mostly consisted of attempts at synthesizing natural sciences and philosophy within religion. Although religion is, of course, an important part of why human societies developed in the directions that they did, religion was also the cause of the initiation of a period of darkness on the planet such as never seen in science before.

THE PERCEPTION OF BIRTH, GENETICS AND HEREDITY DURING THE LONG DARK AGES

Fredrick II, the King of Sicily, had his formal wedding in 1209. But he did not consummate his marriage until the astrologers announced the proper time for producing a male child. At the auspicious time determined by the astrologers, only then did he have intercourse with his bride. He promptly thereafter - and with great confidence - told her that she was pregnant with the heir to the throne. According to the court chronicler, the astrologers had been correct as, nine months later, the queen gave birth to a little prince.

The above story is only one of thousands that existed during the Dark Ages. It was a time of myth and mythology, where human beings believed in legends of unicorns, and myths involving god-like centaurs. Astrologers were considered powerful influencers who could easily determine the outcome of any event on Earth by observing the movement of certain stars. Almost anyone that had status or a reputation to uphold consulted or employed an astrologer.

The medical profession saw some extremes as well, with doctors carrying charts that would indicate how the position of the stars would affect the biological functioning of the human body. They checked the position of the celestial bodies before making a diagnosis and were required by law to consider the position of the moon before conducting a surgical procedure. Such practices often had detrimental effects, as the patient died due to the delay in surgery or medical application.

After the fall of the Greek Empire into the hands of the Romans, subtle changes occurred - yet, very much the same way the Greeks had separated religion from science, the Romans did the same. But things in the Roman world worked a little differently. Rather than focus on natural sciences, they concentrated on the development of advanced technologies, and took engineering and machinery for construction to another level. On top of it all, they had a mighty empire to rule. This resulted in the Romans' focus on utilitarian sciences that were very effective, yet, deprived the society of philosophical thought and advances in natural sciences. In a way, the pursuit of knowledge and critical thought became stagnant after the 5th century.

CICERO'S PURSUIT OF LEARNING & EVALUATING THE ROMAN EMPIRE'S ATTITUDE TO NATURAL SCIENCES, EVEN BEFORE THE DARK AGES

One of the greatest Roman writers - credited with influencing European literature and conceptualizing the idea of liberty - was Cicero.

Marcus Tullius Cicero was born well before the dark ages, on 3 January 106 BC, to a wealthy Roman family living in a hill town known as Arpinium, 100 km from Rome. During this time in history it was required that all wealthy and upper class individuals showcase a level of 'culture.' That meant the young men of wealthy families had to study both Latin and Greek. As such, Cicero studied most of the works of the ancient Greek philosophers, poets and historians, with the theory and practice of rhetoric of the Greek poet, Archias, having the most influence on him. From his knowledge of Greek, Cicero took it upon himself to translate many of the theoretical concepts of Greek philosophers into Latin. His work as a student and astuteness for learning caught the attention of people around Rome, which later afforded him the opportunity to study Roman law under a well-known authority on law, Quintus Mucius Scaevola. Cicero continued to grow and expand his knowledge, gaining higher and higher positions in politics until reaching that of consul in 63 BC.

Roman engineering created archaeological marvels; a large system of roads stretching across the Roman Empire, as well as a phenomenal water supply and sewage system. All of this was accomplished due to the mathematics and geometry of Greek scientists. Cicero wasn't averse to Roman thinking; after all, he enjoyed a comfortable Roman lifestyle. But he did sometimes criticize the Roman way and the lack of knowledge and pursuit of knowledge in the natural sciences. In some of his writings, Cicero comments on the Roman attitude towards Greek natural science, whereby he says: *"We have limited usefulness to this art of measuring and calculating,"* referring to the advances in engineering but not natural sciences. Cicero was concerned that the lack of studies in natural sciences could lead to the decay of human thought and society. He was right, of course; after the Greeks' advances in science, the world experienced centuries of darkness, wherein very little progress was made scientifically. Fortunately, Cicero's work continued to inspire generations of individuals after his death. Thomas Jefferson, one of the founding fathers of the US, often said Cicero's work was one of the major contributors that later influenced the American Independence Movement.

THE ROMAN VIEWS ON NATURAL SCIENCE STUDIES

The Romans did not invest much in natural sciences and philosophy, but they did not discourage their pursuit either. Galen, who wrote the all-too-famous book, "The Book of Medicine", lived for most of his life under Roman rule, but was a Greek himself. (See the previous chapter for more on Galen.) Roman's didn't encourage the pursuit of studies in natural sciences; therefore, their lack of interest had an impact on the subject matter, especially since a large number of nations were, at the time, under Roman rule. As such, natural sciences took a back seat while wars and engineering pursuits enjoyed the most attention.

THE CONCEPT OF HEREDITY IN THE ROMAN EMPIRE AND BEFORE THE LONG DARK AGES

For many centuries, lineages were formed according to the patriarchal line, with many nobles regarding children born of nobles to be noble, while a peasant's child could never become more than a peasant. *"Noble children are born from noble sires,"* was a common belief among the Celts, Greeks, and Romans. But things started changing rapidly, as the Roman Empire expanded and a multi-cultural society formed in the wake of the Romans' military conquests.

Later on, these patriarchal concepts started further evaporating, as people of Germanic races - the Germans, Dutch, Scandinavians and Anglo Saxons to name but a few - didn't strictly organize their societies along patrilineal lines. These races valued both the mother's and the father's contribution to forming a child's character, and considered that kingship came from both parents. They understood the intergenerational transmission of qualities. As an old German folk saying goes, *"You can't make a silk purse out of a sow's ear."*

The Germans' kindred thus consisted of relatives from both sides of the family. Due to these different perspectives, what was known as the 'Roman' way disintegrated and 'new' perspectives and ideas evolved. (The European states that we are familiar with today emerged when the Roman Empire disintegrated.)

North Americans also followed the Germanic concept of a combined patrilineal and matrilineal family.

In Ancient Scandinavia, people believed that an 'inherited luck force' played a role in distributing genes. Kings who did not inherit this force were removed as they were thought to be genetically impure for the kingship.

5TH – 15TH CENTURY AD - CHANGES IN THE CONCEPT OF HEREDITY DURING THE LONG DARK AGES

Before the long Dark Ages, we note that heredity and genetic inheritance were considered more objectively and organized according to the Greek understanding of natural sciences. But things started changing rapidly during the Middle Ages from the 5th century AD onwards.

Theological teachings became the source of understanding, wisdom, and knowledge among many nations around the world. Various religions, such as Christianity and some interpretations of Islam, discouraged natural philosophy and, therefore, strayed from true progress until the 12th century. During this period of the dark ages, cruel persecutions occurred, with thousands slaughtered for what were considered 'heretical' beliefs if they did not follow official religious teachings. This was known as the Medieval Inquisition (1184), which was a series of inquisitions whereby the Catholic Church was charged with the duty to suppress heresy, declaring any beliefs or religions other than Christianity, unlawful. The Episcopal Inquisition followed from 1184 to 1230, and later the Papal Inquisition from 1230 onwards.

Women suffered particularly greatly at the hand of religion, in that they were persecuted for a variety of 'acts' considered ungodly by the Church. A woman who couldn't bear a child to a king was often thought to be cursed, or those who only had female children without producing male heirs, were thought to be 'bewitched.' A male could have a wife and a variety of consorts without prosecution by the Church, while a woman could be put to death for infidelity. Such scientists as did exist had to operate under the utmost secrecy, or face certain death.

THE CONCEPT OF 'ENSOULMENT'

The Soul is the *"immaterial essence, animating principle, or actuating cause of an individual life"* as defined by Webster's dictionary. The true nature of this 'actuating cause of an individual life' has troubled humans since ancient times. Problematic philosophical questions arose, such as: *"What is the difference between a living and a dead body? Is the soul responsible for keeping us 'alive' here on Earth? If I die, where does my soul go?"*

INVESTIGATING THE CHANGING IDEAS ABOUT 'ENSOULMENT'

The ancient Greeks believed in 'Pneuma' (a vital air) entering the body as the baby inhales for the first time. This, they thought, was the difference between a living and a dead body. Some Greek philosophers also believed that it was the process of fertilization that created a human soul. One of the best known Greek philosophers, Aristotle, proposed the theory of progressive ensoulment, where he argued that a human soul developed from a vegetative soul, which became an animal soul before perfecting itself and becoming a human soul.

The early Church also held various views about the soul and when it entered the body. Some thinkers, such as Tertullian, believed that the soul and the body were both created at the same time. However, St. Augustine didn't support this view, simply because it didn't encompass the 'original sin'.

St. Thomas Aquinas and St. Augustine of Hippo did not permit early abortion, despite holding the view that the soul entered the fetus on the 40th day.

14TH THRU 17TH CENTURIES – THE RENAISSANCE

It was around this time that universities in Europe championed the beginning of the Renaissance. The new understanding was that all men were equal in the eyes of the one great creator of the universe. This new wisdom contrasted with the prevailing ideas of genetics, heredity, and, therefore, the hierarchy of talents. The Church didn't recognize the proclaimed *'special features'* carried in the bloodline of kings and noblemen. The Christian God was omnipotent and directed the workings of everything that makes up the universe. The Church propagated the belief that the only way to know and explore nature was to study The Bible. No other writings were allowed to be read or studied.

However, when Christianity and kings clashed, political settlement was achieved by accommodating royalty and giving them 'divine right' to rule as representatives of God, as long as they obeyed The Church. This new 'right to rule by a divine decree' emerged out of a political settlement, rather than inherited bloodline.

IN CONCLUSION

The Dark Ages was a time of suffering and great difficulty for science, as well as for women, with very little tolerance shown for any faith or scientific development other than the religion prescribed.

SOURCES & REFERENCES – CHAPTER 8

Stubbe, Hans and T.R.W. Waters (Trans.), *History of genetics: from prehistoric times to the rediscovery of Mendel's laws.* Cambridge, MA: MIT Press, 1972.

Lindberg, David C., ed. *Science in the Middle Ages.* University of Chicago Press, 1980.

CHAPTER 9:
THE LIGHT AT THE END OF A DARK TUNNEL

" To know that we know what we know, and to know that we do not know what we do not know, that is true knowledge."

- Nicholas Copernicus

Thankfully, the world we create can change. After the long dark ages, scientific thought and observation made a comeback, whereby we started to see glimpses of progress towards human enlightenment and answers to questions about heredity. European universities and scholars started gathering and translating old treatises and manuscripts. The advent of printing allowed knowledge to spread rapidly across the world, facilitating the sharing of ideas and collaboration among scientists. They did not have to redo what others had already accomplished. The decline of the Islamic caliphate and the rise of European countries fueled a flowering of works in science and arts.

This period laid the foundation of the idea that, even though God created life, humans are like animals, except distinguished by the presence of a soul. The body could be studied like a machine. Artists studied human bodies to paint more realistic paintings. Animals were dissected to understand the workings of the body parts. Leonardo Da Vinci, better known for his sculptures and paintings, was, indeed, a scholar of biology. He was one of the first to make a connection between animals and humans in far more detail than Linnaeus did in classifying humans.

A careful study and dissection of animal organs automatically led to an understanding of their functions and inter relationship with one another. Plants were studied for their medicinal value. Not only were the advances in physical sciences and astronomy spectacular, like the recognition that the earth revolved around the sun rather than the other way around, but also significant progress was made in biology.

PLANT REPRODUCTION AND TRANSMISSION OF CHARACTERISTICS.

Babylonians, and later Arabs, practiced artificial fertilization of date palms based on empirical knowledge. However, this astute observation was not advanced to explore the possibility of sexual reproduction in plants, which was considered to be exclusive to the animal kingdom. Europeans only learnt about the sexual reproductive phenomenon in plants in the late seventeenth century.

In 1694, Rudolf Jakob Camerarius, a German botanist and physician, published his work, which clearly showed that pollen was the male component and the seed-bearing part the female equivalent in plants. A union of these two was a necessary prerequisite for the production of a new plant offspring. Soon thereafter, this knowledge was applied to explore artificial cross-pollination and hybridization. It was easy; unlike animals, that could not be forced to mate with different species, pollen from one plant could be easily transferred to the female part of another plant. Josef Gottlieb Koelreuter, another German botanist and physician, can be described as the father of plant hybridization; his curiosity and meticulous experimentation produced the first well-documented plant hybrid. His work was further expanded by others, ultimately leading to the seminal work of Mendel, covered in a later chapter.

Scholars in seventeenth century Europe organized societies to work for the advancement of knowledge. Governments later founded National Academies to promote scientific scholarship and international cooperation. These societies and academies instituted prize competitions to explore scientific questions of the time. These prizes spurred a flurry of scientific exploration and experimentation, advancing science and scholarship.

WILLIAM HARVEY'S THEORY OF EPIGENESIS AND THE CHANGES OF THE CONCEPT OF HEREDITY AFTER THE DARK AGES

In the 17th century, an English physician named William Harvey made an extensive study of the uterus and development of organisms. He proposed the theory of *epigenesis*. According to this theory, a homogenous material forms a functional fetus that undergoes various gradual changes to develop into an organism. He showed that the fertilization of the egg by the sperm leads to the development of life.

His studies were in stark contradiction to the popular notion of preformation, whereby it was thought that the sperm of a man or the egg of a woman already had an unfolded organism within it that slowly developed into a full organism. The tiny organism was called a *homunculus*. The Church supported this idea of preformation, as it believed that 'life that exists on Earth was created at the moment of creation.' This idea, similar to Russian nesting dolls, appealed to Christians who thought all humans were born sinners, and that they were already present at the time of 'creation', which was thought to have happened about 6,000 years ago. Luckily for science, this theory was abandoned with the development of the microscope. Today, we know that the biological material for the child comes from both the male and female, while the male's sperm is the determining factor as to the gender of the child.

IN THE SHADOWS

Until the 20th century, the birth and sex of a child remained a shadowed area, covered in the fog of ancient Greek knowledge and outdated investigations into the development of life. Many claimed that astrological forces, a woman's 'thoughts' during intercourse, and certain foods could influence a child's sex. Second-century Greek physician, Galen, remained an authority on determining sex. However, he wasn't the only one influencing doctors and authors; nearly five hundred different theories on this subject existed by the nineteenth century, as pointed out by Patrick Geddes and J.A. Thomson in 1889.

It was in only in 1916 that an American biologist, Calvin Bridges, discovered that male sperm carried the X or the Y-chromosomes. Contrary to what humans believed about the determination of sex from the ancient Greek period to the nineteenth century, this discovery showed that it was the father's sperm cell that played the decisive role in sex determination. A combination of X and X from both parents leads to a female child and a combination of X and Y to a male. As only the male can contribute the Y chromosome, it's the father's sperm which determines the gender of the baby.

IN CONCLUSION

Christian theology clearly influenced philosophical thinking in Europe for centuries. The 12th century Renaissance movement sparked considerable creative energy throughout Europe, encouraging scientists to again investigate the phenomena of nature using careful scientific method. Scientists like Galileo and Copernicus made observations using sophisticated scientific devices that changed the centuries-old Aristotelian and Biblical view of Earth as the center of the universe. Although many scientists had to operate in secret during the Dark Ages, there were scientific discoveries that would later set the stage for modern advancements.

SOURCES & REFERENCES – CHAPTER 9

Jardine, Lisa. *Ingenious pursuits: Building the scientific revolution*. Anchor, 2000.

Mullins, Lisa. *Science in the Renaissance*. Crabtree, 2009

CHAPTER 10:
MAN - CREATED IN GOD'S IMAGE OR A BRANCH OF THE MAJESTIC TREE OF LIFE?

"Nature reveals her secrets only to those who wish to discover the truth."

- Dirk Dunbar

Carl Linnaeus, known as the man who named everything in the plant and animal kingdom via a binomial system, was born on 23 May 1707. His father, Nils Linnaeus was a clergyman and his mother's name was Christina. They used to hang flowers on their child's bed, and even noted that whenever Carl was restless, he would become calm if he had a flower in his hand.

As a child, Carl often got confused with the long scientific names of plants and found it impossible to remember them all. As the story goes, Carl had access to vast meadows, marshland, and a well-stocked garden as a kid. He wanted to remember the names of everything that was around him. But it wasn't easy, especially when the scientific name for tomato had eight words in it. He would often forget the names of plants and ask his father repeatedly, only to add to his frustration. As he grew older, Carl took it as a challenge to simplify the world of naming animals and plants and set about a lifelong quest to remember (and later organize) the names of plants and animals.

Thanks to Carl, the name for tomato that he used to struggle with, namely, *Solanum caule inermi herbacio folis pinnatis recemis simplcibus*, later became *Solanum lycopersicum*.

CARL UNDERSTOOD SEXUAL REPRODUCTION IN PLANTS FROM A YOUNG AGE

"The flowers' leaves serve as bridal beds which the Creator had so gloriously arranged, adorned with such noble bed curtains and perfumed with so many soft scents that the bridegroom with his bride might there celebrate their nuptials with so much the greater solemnity." (Linnaeus, 1707 - 1778)

Since Carl's father was a clergyman and most children followed in the footsteps of their parents in those days, as a child, Carl expected to become a priest. He attended school from 6 am to 5 pm daily, where he learned various subjects but never became a distinguished student. Unfortunately, Carl wasn't recommended for the priesthood by the school. Instead, a renowned physician, Dr. Johan Rothman, gave him a chance as his apprentice.

Dr. Johan helped Carl explore the world of plants through herbal medicine. It was during this time that he learned about the sexuality of plants, which later inspired him to achieve greater things. While working with Dr. Johan, Carl understood that plants and humans shared similarities in the way they sexually reproduce. He was fascinated by flowers and used them for grouping plants into different categories, relying on the number and arrangement of stamens and stigmas, which are the sexual parts of a flower.

CARL AT THE UNIVERSITY OF LUND

Carl was born to poor parents, who couldn't afford to send him to a medical school. However, Carl had a passion for medicine and managed to find a way into the University of Lund where he excelled and inspired influential people around him.

Dr. Kilian Stobaeus, who was a famous physician of the time, arranged for his accommodation, allowed Carl access to his lectures and an extensive collection of books. The doctor also showed his collection of dried plants to the willing young student. Carl, having never seen such a thing before in his life, was fascinated and decided to develop his own collection of dried plants, which became internationally famous later in his life.

CARL LINNAEUS AT UPPSALA UNIVERSITY – THE DEVELOPMENT OF THE BINOMIAL SYSTEM

Carl continued his medical education at Uppsala University, where he met and was influenced by a learned naturalist, Olaf Celsius. Carl became a teacher at the University and, in turn, inspired many of his students to collect plants. At the tender age of 24, Linnaeus had achieved a lot.

At Uppsala University, Carl Linnaeus met Pehr Artedi. Both shared an interest in studying the natural world around them, so they set out to develop the method for naming organisms and describing them properly. Artedi studied fishes, amphibians and reptiles, while Carl decided to spend his energies exploring insects, birds, and plants.

Artedi died prematurely, but Carl continued their work and developed the binomial system of naming plants and animals. It consists of two names for each living organism. He traveled to the coast of Sweden, Norway, and Finland to study the flora of the area, and in 1737 published his work under the title, "*Flora Lapponica*," documenting the Arctic and Alpine species.

UNDERSTANDING WHAT THE BINOMIAL SYSTEM IS

According to Carl's system of classification, all living organisms belong to either the animal kingdom or the plant kingdom. The animal kingdom has two further classes named vertebrates and invertebrates, as well as groups and subgroups based on their common characteristics. Using this approach, scientists today place human beings in the same subgroup as chimps and apes. It asserted that man wasn't a special creature but an animal that shared many characteristics with other living organisms.

Although Carl had strong religious beliefs and never considered evolution as a reason for such a variety among living organisms, his theological school of thought focused on using reason and observation to prove the existence of God and his creation.

Carl believed that one could understand God by studying his creation. His famous words, *"the earth's creation is the glory of God,"* clearly indicate that he tasked himself to create a natural classification of the world around him to understand the order of the universe.

THE LEGACY OF CARL LINNAEUS

Linnaeus used this two-name system to help remember all the plants while collecting and categorizing them. He probably never thought it would later become the standard used far and wide across the world. But, the Carl Linnaeus binomial system became his legacy and the reason he is often referred to as 'the man who named everything.'

Overall, Carl Linnaeus named a total of 14,000 species; with his system helping us to name over 1.5 million species today. His classification not only standardized the system of naming a biological species but also a way to place them in an order which showed how closely related they were to each other. He chose the name Homo sapiens for humans. He also placed orangutans and chimpanzees, the only two apes known at the time, in the same genus as humans, recognizing that they were closely related to each other. Although he believed that God created humans as special beings, his careful observations could see the connections between different living beings.

This systematic approach to analyzing the features of so many species and placing them in an orderly pattern established that all species were related to each other. This concept further evolved over time, as the evolution of species from the simplest forms to more complex ones. He is a celebrated hero in his homeland, Sweden, where he is one of the few scientists whose picture appears on currency notes.

IN CONCLUSION

Carl Linnaeus was the greatest biological name-giver. His method of classification of the plant and animal kingdom helped us name over 1.5 million species. Even though he didn't propose the theory of evolution, Charles Darwin studied Carl's classification system and made observations that led to the theory of evolution. The Biblical and mythological concept of man at the center of the universe and created in the image of God was now presented as simply another branch in the majestic tree of life.

SOURCES & REFERENCES – CHAPTER 10

UCMP Berkeley. *Carl Linnaeus (1707-1778).* [Online] Available at:
http://www.ucmp.berkeley.edu/history/linnaeus.html

Carl Linnaeus- The Linnean society. https://www.linnean.org

Carolus Linnaeus. Famous Scientists: The art of genius. [Online] Available at:
https://www.famousscientists.org/carolus-linnaeus

CHAPTER 11:
THE INVISIBLE COMES TO LIFE

"By the help of microscopes, there is nothing so small, as to escape our inquiry: hence there is a new visible world discovered to the understanding."

- Robert Hooke

Henry Oldenburg, the secretary of the Royal Society of London, received a strange letter in 1673, from a young lens maker who described dental plaque in intriguing detail from looking at it closely through a microscope. It was the first-time humans had observed and documented microbiological life.

The amateur lens maker was Antoni Leeuwenhoek. He not only studied microbial life in the plaque between his teeth but also took samples from a variety of people around him.

LIVING ANIMALCULES THROUGH ANTONI'S MICROSCOPE

"I then most always saw, with great wonder, that in the said matter there were many very little living animalcules, very prettily a-moving."

This is how Antoni, the Dutch lens-maker, first described the microscopic world he saw in his own dental plaque. He carefully described all the details, such as how the organisms moved around in the stuffy material and how they differed from each other. In his own words:

> *"The biggest sort... had a very strong and swift motion and shot through the water like a pike does through the water. The second sort... oft-times spun around like a top..."*

He also noticed that these organisms were numerous. When he looked at the samples of an old man who had never brushed his teeth, he was simply baffled. He described the vast variety of tiny organisms in these famous words:

> *"An unbelievably great company of living animalcules, a-swimming more nimbly than any I had ever seen up to this time."*

THE HISTORY OF MICROSCOPES

The invention of the microscope changed how we looked at our world. It started the field of microbiology and revolutionized biology. Antoni Leeuwenhoek made the first documented observations.

The Romans, from the first century AD, were the first people to take advantage of the magnification properties of glass. They had 'orbits', made of glass and filled with water, that helped them to magnify objects.

In the thirteenth century, Italian lens-makers were already selling eyeglasses. Alessandro Della Spina of Pisa was the first documented eyeglass maker.

The earliest microscopes only had one lens. They were essentially used to magnify insects, while traders used them to look closely at fabric to determine its age and quality.

In the 1590s, Dutch lens-maker, Hans Jansen and his son developed the first compound microscope. Jansen observed that putting several lenses in a tube dramatically magnified the object at the other end.

These early microscopes baffled everyone, but they didn't provide a clear enough view of the tiny world of microbes, and therefore, were rarely used in scientific study.

Robert Hook (1635-1703) was the first noted biologist to extensively study the microscopic world. His book, *'Micrographia,'* features a large number of hand-drawn illustrations. Robert Hook's work inspired many. Among them was Leeuwenhoek, who made many improvements to the compound microscope that Hook used, converting it into a highly sophisticated scientific tool.

LEEUWENHOEK'S EARLY LIFE AND HIS JOURNEY TO BECOME THE GREATEST LENS-MAKER

Van Leeuwenhoek was born in 1632, in the city of Delft, to a basket maker. He attended elementary school in Warmond and then went on to live with his uncle, who taught him physics. At the age of sixteen, he went to Amsterdam for an apprenticeship at a linen-drapers shop. When he returned home after spending five years in Amsterdam, he opened his own shop and made enough money to pursue his passion. It was here that Leeuwenhoek met cloth merchants and saw their magnifiers that used a low-power magnifying glass. At the age of 36, Leeuwenhoek finally started making instruments that would take magnification to the next level.

ANTONI LEEUWENHOEK'S MICROSCOPES

Leeuwenhoek used simple techniques to develop his microscopes. He would take soda-lime glass bottles and a hot flame to make small, but high-quality glass spheres. These spheres were used as lenses in each microscope.

Leeuwenhoek's microscopes were three to four inches long, with short focal length. The objects had to be placed extremely close to the eye. People with poor eyesight found it impossible to see anything using Leeuwenhoek's microscopes and they were useless without proper lighting conditions.

These magnifying tools contained a single lens that was mounted between two copper plates. They used threaded screws to adjust the zoom and the object being observed. Leeuwenhoek was also able to rotate his microscopes using the same threaded screws that he used to adjust the focus.

Compound microscopes had been in use for nearly a century, but Leeuwenhoek significantly improved on them. They had a magnification power of around 30X, which was nowhere close to Leeuwenhoek's single lens microscopes that magnified to 270 times the actual size.

The Dutch lens-maker was able to see objects as small as 0.004 of an inch with his carefully developed magnifiers. It made him curious; therefore, he started looking at everything around him. He examined water and saw living creatures in it. He looked at sperm cells and blood corpuscles. Leuwenhoek also discovered bacteria, which he called animalcules. To record his observations, the Dutch scientist hired an illustrator.

LEEUWENHOEK'S CORRESPONDENCE WITH THE ROYAL SOCIETY

Unfortunately, Leeuwenhoek didn't have the opportunity to learn Latin. It was impossible to communicate with other scientists and explore their work without knowing Latin, which limited Leeuwenhoek's observations only to his friends.

It was Rognier de Graaf, a young physician and a friend of Leeuwenhoek, who actually wrote the famous letter to Henry Oldenburg in 1673 and explained what Leeuwenhoek had discovered. Since the Royal Society responded to the letter, Leeuwenhoek was able to establish official correspondence with the Royal Society.

In the years that followed, the Dutch scientist shared everything that he had observed with his microscope in nearly 500 scientific letters.

While recording his observations and documenting them in his letters to the Royal Society, Leeuwenhoek took great care. He used a special scale that he devised himself to measure the tiny organisms. His measurements were actually so accurate that they are very close to modern measurements of the microscopic world.

Leeuwenhoek's discoveries are numerous. He studied the giardia parasite, along with many other previously unknown bacteria. He took samples from everything around him and observed microscopic life found in each sample. He noted that an area the size of a sand grain could contain over 1,000 microorganisms. He also observed and compared red blood cells in many different species.

THE THEORY OF SPONTANEOUS GENERATION AND LEEUWENHOEK'S LEGACY

Aristotle proposed that a substance called *'pneuma'* was responsible for giving life to maggots that appeared on dead animals. The great philosopher believed that this life-giving substance was present all around us. Since people were unable to see the microscopic world, they found it convenient to believe this theory of spontaneous generation. But Leeuwenhoek's discoveries of the microscopic world allowed people to challenge the conventional wisdom of spontaneous generation, as we will see in the next chapter.

Van Leeuwenhoek was elected a *Fellow of the Royal Society* in 1680, in recognition of his discoveries in the field of microbiology. Even though he was ignorant of Latin and Greek so couldn't read the classical texts, and had never attended a university, he became an outstanding scientist. It was his curiosity and his dedication to his world that helped overcome his lack of education.

In his own words:
> *"Some go to make money out of science, or to get a reputation in the learned world. But in the lens grinding and discovering things hidden from our sight, these count for naught. And I am satisfied too that not one man in the thousand is capable of such study, because it needs much time and*

you must always keep thinking about these things if you are to get any results. And over and above all, most men are not curious to know: nay, some even make no bones about saying, what does it matter whether we know this or not?"

Leeuwenhoek also talked about his motivation; a craving for knowledge, and not praise and money. He enjoyed knowing the unknown and finding remarkable things around him. He carefully documented his every observation to ensure that he could pass them on to the generations to come.

Leeuwenhoek died in 1723. Unfortunately, no one picked up his studies; therefore, after his death, progress in microbiology slowed down because of the limited availability of microscopes and the lack of interest in microbes among scientists. The microscope remained nearly the same for another century.

AN UNFORTUNATE LACK OF INTEREST AFTER VAN LEEUWENHOEK'S DEATH

One of the reasons microbiology saw a lack of interest after Leeuwenhoek's death was that scientists had not yet understood the importance of microscopic organisms and their role in biology. Many learned authors continued to believe that life could arise from nothing.

Thankfully, Luis Pasteur changed this notion in 1862, with his famous experiments, and encouraged scientists to develop better microscopes to study the unseen world closely.

With time, microscopes became more powerful, enabling scientists to watch cells and bacteria closely and identify their structure and movement. In 1838, two German scientists, named Mathias Schleiden and Theodor Schwann, studied cells and found out that they are the basic building blocks of living things. Similar observations were made later by Rudolf Virchow, leading to the famous cell theory.

IN CONCLUSION

Today, only a few of Leeuwenhoek's microscopes have survived, although he made more than two hundred of these magnifying tools. However, his incomparable contribution to the field of microbiology will survive for centuries to come.

SOURCES & REFERENCES – CHAPTER 11

Dobell, Clifford. "Antony van Leeuwenhoek and his "Little animals": Collected, Translated and Edited from His Printed Works, Unpublished Manuscripts and Contemporary Records". *Journal of American Medical Association* 1933:100(5)363.

CHAPTER 12:
LIFE BEGETS LIFE: CELL THEORY

"Omne vivum ex vivo"- All life from life.

- Louis Pasteur

Humans have always been curious about what things around them are made of and what purpose they serve in the grander scheme of things. Dalton, in 1810, had formulated that all matter was made up of tiny units called atoms. Could living things be similarly made from some fundamental units, common to all life forms? From Greek philosophers to Saint Augustine of Hippo, thinkers had all tried to provide the answer for this puzzle.

Theodor Schwann decisively provided a comprehensive answer to this troubling question for the first time in 1847, describing the 'cell' as the basic structural unit for all living organisms. Known as Cell Theory, it proposed that *'tissues are formed of cells in an analogous, though diversified manner'* and the formation of all things is governed by *'one universal principle of development'*, which starts with *'formation of cells'*.

EXISTING VIEWS ABOUT THE FORMATION OF LIVING THINGS

SAINT AUGUSTINE OF HIPPO'S VIEW OF DIVINE DECREE

Saint Augustine of Hippo was probably the most influential theologian in the history of the Christian Church. In one of his books, he cited a passage from The Bible to establish the principle of 'ongoing creation'. The passage reads:

"Let the waters bring forth abundantly the moving creatures that hath life." (Genesis 1:20)

He further relied on these Biblical commands: *"Let the land produce vegetation. Let the water teem with living creatures."*

According to St. Augustine's principle of ongoing creation, the divine decree was issued at the time of creation and it is what continues to give life, creating creatures from earth and other materials.

St. Augustine was an influential writer, so others built on his concept of divine decree. Some naturalists wrote bizarre stories, mainly a work of their imagination under the influence of the concept of divine decree. A naturalist from the 12th century, Alexander Neckham, believed he had seen geese sprouting from the resin of a pine tree. Since these birds don't appear in Scotland and Ireland during the summer, he speculated that they may spontaneously generate from nonliving materials. His findings read:

"They are produced from the fir timber tossed along the seas and are at

first like gum. They hang down by their beaks as if they were a seaweed attached to the timber, and are surrounded by shells in order to grow more freely."

This incredibly incorrect account of geese coming to Scotland from the North during the winter was accepted and remained unchallenged for years.

ARISTOTLE'S THEORY OF SPONTANEOUS GENERATION – LIVING THINGS COULD COME FROM NON-LIVING THINGS

Aristotle, the father of Western natural philosophy, wrote about the nature of things around us in his famous book on the history of animals. He observed that some animals spring from their parents, while others 'grow spontaneously and not from kindred stock'. This theory of spontaneous generation remained in force for the next two millennia.

Aristotle saw maggots rising from dead flesh and fleas from dust. He couldn't understand what was actually happening, as he could only see with his naked eyes. Based on his observations, he theorized that a *'vital heat'* is present everywhere, including in non-living things. This *'vital heat'* gives life and he considered it to be the reason that living things could come from non-living things.

It was during the Renaissance period that Aristotle's theory of spontaneous generation was finally discarded. Before that time, it was considered a 'fact'. Even Shakespeare endorsed the idea of spontaneous generation as he famously said in a passage:

"Your serpent of Egypt is bred now of your (Nile) mud by the operation of your sun: as well as a crocodile."

REDI'S FAMOUS EXPERIMENTS - THE FIRST EVIDENCE AGAINST SPONTANEOUS GENERATION

It was only in the year 1688 that Francesco Redi conducted experiments to challenge the theory of spontaneous generation proposed by Aristotle centuries ago. In his famous experiments, he took a piece of meat and placed it in different containers. Some of these containers were sealed, while others were open. Some were even partially covered.

The containers that were properly sealed didn't have any worms, while the open containers had worms after a few days, as flies visited the meat and laid eggs. Redi himself believed in the theory of pre-existence that asserts that divine force created all living things and they undergo subsequent generation through their parents.

In 1859, the French Academy of Sciences announced prize money of 2,500 Francs for anyone who could give evidence for or against spontaneous generation. As a result, the famous French scientist, Louis Pasteur, received the prize in 1862 when he famously showed that living things came from living things. In his experiment, he took two glass flasks and placed nutrient solution in both of them. He bent the neck of these flasks into an S shape. He then sterilized both the flasks and broke the neck of one, exposing the nutrient material inside to the air. The other flask had its neck intact, which kept the solution away from any outside influence.

Surprisingly, the solution in the broken flask became cloudy, while the other remained clear. It showed that it was actually the multiplication of microorganisms that came through the air and not life arising from nonliving material.

THE INVENTION OF THE MICROSCOPE AND THE ORIGINS OF CELL THEORY

The invention of the microscope offered a new perspective on the unknown and unseen world. With this sophisticated instrument, scientists were able to look at microorganisms and plants, and see for themselves how these were made of cells. According to Mathias Jacob Schleiden:

> *"Every plant developed in any higher degree is an aggregate of fully individualized, independent, separate beings, even the cells themselves. Each cell leads a double life: and independent of, pertaining towards own development alone, and another incidental, in so far as it has become an integral part of the plant."*

Even before scientists could see the cell with a microscope, philosophers had speculated on the idea of smaller constituent parts of the human body, invisible to the naked eye. The most popular idea was the one proposed by the French physician, Marie Bichat, who divided the human body into twenty-one different kinds of tissues. She considered each of these tissues as part of the fundamental structure.

SCHLEIDEN AND SCHWAN

Schleiden was born in Germany in 1804. He studied law but practiced only for two years. Later, he studied medicine and botany and used the microscope to look at plants. During his time, the microscope had become a very sophisticated tool of scientific observation with two lenses with a wide aperture range.

His colleague, Schwan, was born in Germany in 1810. He studied medicine and pursued a career as a professor at a university in Belgium. Unlike Schleiden, Schwan was more interested in studying animals. Chick embryos and tadpoles were among his favorite study subjects under the microscope.

For the first time in history, these advanced microscopes helped Schleiden and Schwan look closely at the structure of plants and animals. After several observations, they concluded that:

> *"The elementary parts of all tissues are formed of cells in an analogous, though diverse manner, so that it may be asserted, that there is one universal principle of development for the elementary parts of organisms, however different, and that this principle is the formation of cells."*

SCHWAN AND SCHLEIDEN WERE NOT THE FIRST ONES TO TALK ABOUT 'CELLS' AS FUNDAMENTAL BUILDING BLOCKS IN LIVING ORGANISMS

It was Schwan and Schleiden's work that developed the concept of cell theory and gave it universal acceptance. However, many scientists had already made similar observations, including Robert Hooke, Leeuwenhoek and Malpighi. Nevertheless, it was after Schwan's paper that cells were regarded as the fundamental building blocks of all living things, including plants and animals.

But the credit should probably go to Rene Dutrochet for studying the previous data and making his own observations that led to an understanding of how living organisms are structured.

WHO WAS RENE DUTROCHET?

Rene Dutrochet was born in France in 1776. He had a birth defect in his foot that seemed impossible to fix at the time. But his mother continued searching for a treatment until she found someone who treated Rene's foot successfully.

In 1806, Dutrochet obtained a degree in medicine. In his famous thesis, he explained the mechanism of the human voice. He joined the army as a surgeon but retired early due to a severe attack of typhoid. When he returned home, he set up a small home laboratory to continue his research work. During this time, he wrote many papers and sent a majority of them to the Paris Academy of Sciences.

DUTROCHET'S WORK AND HIS CONTRIBUTION TO PLANT PHYSIOLOGY

Rene made some remarkable contributions to the field of plant physiology, and in recognition of his work he was chosen as a member of the Paris Academy of Sciences. His scientific investigations in the field of histology and embryology are considered his most important works. Along with studying the structure of cells, Rene also looked at how they would swell - a process known as osmosis. From his various observations, he concluded:

> *"Physiological connections which I have established between plants and animals make it clear that there is but a single physiology, a general science dealing with the functions of living beings, functions which vary on the mode of execution but which are fundamentally identical in all organized beings."*

It is unclear whether Schleiden and Schwann read Rene's papers. However, Schleiden did mention Rene in a footnote in his paper, indicating that he had already read Rene's research.

Whether the credit for cell theory goes to Schwan and Schleiden or Rene is probably not that significant, since their studies didn't include an important part of modern cell theory, which states that cells arise from pre-existing cells. According to Schleiden, new cells were formed through a process called spontaneous crystallization. According to him, the cytoblast (now known as the nucleus) forms a transparent layer around it once it has reached its full size. It then crystalizes in the presence of a liquid containing sugar, gum, and mucus. He called this material *cytoblastema*.

MODERN CELL THEORY AND VIRCHOW'S ARCHIVES

Modern cell theory has three important components.

1. The cell is the fundamental unit of all organisms;
2. All organisms comprise of cells, one or more;
3. All cells come from pre-existing cells through cell division.

It was German Doctor and pathologist, Rudolph Virchow, who observed the important phenomenon of cell division. Virchow proposed that the cell was the basic unit for all living organisms. Moreover, he showed that cells arise from the division of existing cells. Virchow was convinced that humanity would make massive progress in the field of medicine once they studied the cell more carefully, using clinical observation, experiments, and microscopic observations.

Rudolph Virchow started a new journal named, 'Virchow's Archives.' He published only tested and original work in this journal and made incredible contributions to the field of cell biology. Virchow also studied Robert Remak's work, which had used different mixtures to harden the cell wall and proved that 'some' cells arise from the division of existing cells. Unfortunately, Remak didn't receive attention for his discoveries until Virchow published them in his famous journal as his own work.

IN CONCLUSION

The understanding of what made up the animal and plant world answered a lot of scientific and philosophical questions. It established that organisms are created in the same fundamental way, known as cell division. Their activities are the sum of activities of their cells, whether one or more.

The development of cell theory helped us to understand the relations and interactions of different cells in a living body, giving a new direction to biology and the field of medicine. It is not far-fetched to say that cell theory had a similar influence on scientific progress, to that of the atomic theory of matter.

SOURCES & REFERENCES – CHAPTER 12

Cohen, Martina. *What Is Cell Theory*? Crabtree Publishing Company, New York N.Y USA 2010

CHAPTER 13:

IF MAN CAN DO IT, WHY NOT NATURE? DARWIN: EVOLUTION BY NATURAL SELECTION

"Most of our ancestors were not perfect ladies and gentlemen. The majority of them weren't even mammals."

- Robert Anton Wilson.

In 1760, Robert Blackwell took over the family farm after his father passed away; beforehand, he had traveled Europe and studied farming methods. On their farm, the cattle were left unchecked and randomly procreated, which was a common practice at the time. Blackwell began to control the procreation of his cattle by separating the males and females, then selectively breeding the animals with characteristics that he required. The results were that the old Lincolnshire sheep became the new Leicester sheep, giving better meat and wool. Today, Blackwell's method is common practice in modern farming.

The question thus arose: *"How does experimental breeding help us understand natural selection?"* Before we can answer this, it is important to understand what natural selection is and where the concept came from. Let's have a look:

CHARLES DARWIN

Darwin came from wealth and knowledge; his father was a doctor and his grandfather a known botanist. Darwin set out to become a physician in accordance with his father's wishes, but could not stand the sight of blood; thus, he chose his own path: the natural sciences. He enrolled at Cambridge University and, while there, was recommended by his mentor for a position on the H.M.S Beagle, a 90 ft. British Navy survey ship, about to travel around the world - as an unpaid naturalist. Before the departure, Darwin went to work getting advice and tools for the journey. The captain wanted someone competent enough to realize all the advantages of this scientific trip, so Darwin was the perfect candidate.

THE MOST IMPORTANT SCIENTIFIC VOYAGE IN HISTORY

On December 27th, 1831, the H.M.S. Beagle set sail from Britain to complete its intended two-year voyage, but the voyage continued for five years. During this time, Darwin studied a large variety of fauna and flora all over the world, particularly down the coast of South America. He made numerous notes and sent home many specimens for later study; of the geology, animals, plants and people of all the places that the ship visited. But it was on the Galapagos Islands that Darwin found something significant. He observed that every island's plants and animals differed from the next, but was it possible that they ascended from the same species?

DARWIN'S FINCHES

The now mature scientist became a prolific writer and wrote about the voyage, publishing a history of the journey, as well as a five-volume book of zoology. While studying the specimens he collected, he came across something interesting; with the help of an ornithologist, Darwin identified thirteen (13) species of finches collected from the Galapagos Islands. On the mainland of South America, Darwin had knowledge of only one species on the large landmass, but 600 miles east lay the Galapagos with their thirteen different species; and that too, on small islands so close together.

Darwin found that the main differences between all these species were the shape and size of their beaks. He noted not only the fauna and flora, but also the environments these birds lived in and the food they ate. Though the islands are close together, their climates were slightly different; on the waterless islands, the birds were more suited to eat from the cacti, while on the islands that had copious flowers, the birds were suited better for drinking nectar, whilst still others had beaks to crack the shells of seeds.

During that time, French Biologist Jean-Baptiste Lamarck theorized that the environment changes the features of species to suit the surroundings, which are then conveyed to the following generation. Darwin doubted this concept, as he knew that:

> [...] *Jewish boys circumcised at birth - an acquired change - do not father children with circumcised penis at birth.*

Having this in mind, Darwin realized that the finches that originated on the mainland of South America had slightly different sized and shaped beaks. From there, they then spread across the islands. As they spread over the islands, they needed to adjust to their new environments, changing physically over generations so that they could finally persist and procreate. Some of the finches that came from South America had the right size and shape of beak for the island they came to, and only they survived and flourished, while the finches that could not adjust, died out. Eventually, the thriving species became the only species left on a particular island.

Darwin then realized that, in any populace, the organisms differ from one another and that those with the features that are best suited for the environment thrived, while those without those features perished. Thus, the environment decides which species thrives and which does not. Today, this concept is known as the process of natural selection.

THE PROCESS OF NATURAL SELECTION

In 1838, Darwin picked up a book by the English economist, Thomas Malthus, wherein Malthus states that the human populace will grow two-fold every 25 years if it is not restricted. We believe that this information led Darwin to the realization that this also relates to the plant and animal populaces.

"A struggle for existence inevitably follows from the high rate at which all organic beings tend to increase. Every being, which during its natural lifetime produces several eggs or seeds, must suffer destruction during some period of its life, and during some season or occasional year, otherwise, on the principle of geometrical increase, its numbers would quickly become so inordinately great that no country could support the product." (Darwin, 1859)

This is now known as 'Survival of the Fittest,' but Darwin did not at first publish the notion of natural selection, or survival of the fittest. He first meticulously studied his collected data and performed numerous breeding experiments, as the 1830's and 1840's were the pinnacle of faith in Christian doctrine in England. Darwin knew that his theory, that challenges the biblical account of creation, would be seen as blasphemy. It was only in 1844 that Darwin wrote a 230-word text about his notion of evolution, which was only given to a few British scientists.

CHARLES DARWIN AND ALFRED WALLACE

Wallace was a young British naturalist, collecting plants and animals in Southeast Asia to sell to museums and collectors. In 1858, he sent a draft of his essay, "*On the tendency of varieties to depart indefinitely from the original type*", to Darwin, wherein he had made the same assumption as Darwin. Darwin swiftly completed and published his 460-page copy titled, "*On the Origin Of Species*", which was sold out the day it was published. In the following 20 years, Darwin wrote 15 books, examining his theory further with examples. Wallace's name was on the first scientific presentation of Darwin's paper on the subject in July 1858, but it was Darwin's book that would, to this day, be considered one of the most significant scientific writings; an ideal combination of meticulous notes, preservation and logical thinking. Of course, the theory of natural selection took some time to generate acceptance in the world, as many thought the theory was going against their theological beliefs.

PANGENESIS AND GEMMULES

At first, Darwin could not plainly explain how new species came to be, or how features were passed down from parent to child. So, in 1868, he suggested a theory that he called 'pangenesis'. With no new evidence, Darwin proposed that every cell in the body gives off minuscule elements of legacy; Darwin dubbed them 'gemmules'. Gemmules gathered in the gonadal cells, through which the progenies then gained them. Darwin supposed that an environment could cause alterations in the gemmules, through which these alterations are then given to the progenies.

However, Darwin came to the realization that his theory was not comprehensive, though it would endure as such until his 'gemmule', the unit of heredity, the gene was found and confirmed. However, contrary to the views of Darwin and Lamarck, the hereditary unit is always present in all the cells, and not contributed by different parts of the body to the gonadal cells. His theory also did not identify the source of the mutation that allowed natural selection to function.

PROOF IS FOUND

Hugo de Vries, a young Dutch botanist, found Darwin's book interesting, to the point where he visited Darwin in England. Hugo de Vries had a doctorate in plant physiology and investigated plants comprehensively. He found that some plants exhibited unexpected changes, which he called mutations; he further suggested that these sudden mutations were the cause of evolution. Changes or mutations that provided an advantage to the new species allowed them to flourish, as they would have a survival advantage.

IN CONCLUSION

Darwin, after realizing his curiosity for nature, found that plants and animals with features to suit their environments survive. This realization could explain why some plants and animals were more abundant in certain areas than in others.

At that time, though, Darwin had no way to prove his theory, but was set to tell the world; thus, through his perseverance he found likeminded scientists who could help him - although it took years to prove his theory. Today, natural selection is a proven concept, as well as Darwin's theory of evolution by natural selection.

> *"Nothing at first can appear more difficult to believe than that the more complex organs and instincts should have been perfected not by means superior to, though analogous with, human reason, but by the accumulation of innumerable slight variations, each good for the individual possessor. Nevertheless, this difficulty, though appearing to our imagination insuperably great, cannot be considered real if we admit the following propositions, namely, – that gradations in the perfection of any organ or instinct, which we may consider, either do now exist or could have existed, each good of its kind, – that all organs and instincts are, in ever so slight a degree, variable, -- and, lastly, that there is a struggle for existence leading to the preservation of each profitable deviation of structure or instinct. The truth of these propositions cannot, I think, be disputed."* (Darwin, 1859)

Today, we use this concept not only to strengthen cattle, but also to change the molecular structures of crops. GMO's, genetically modified organisms, are essentially working proof of what Darwin discovered about evolution.

SOURCES & REFERENCES – CHAPTER 13

Darwin, C. 1809-1882. *On The Origin of Species by Means of Natural Selection*. London: John Murray, 1859. [Online] http://www.robmacdougall.org/1805/h1805-18-origin-of-species.pdf

CHAPTER 14:
MONASTERY TURNED SCIENCE LAB:
MENDEL, THE FATHER OF GENETICS

"I am convinced that it will not be long before the whole world acknowledges the results of my work."

- Gregor Mendel

During the 19th century, Charles Darwin wrote and published his most famous book, *'On the Origin of Species,'* which became extremely famous and the cause of hefty debates between religion and science. However, his theory of evolution had one major flaw; the theory proposed natural selection and transmission of qualities, but Darwin had no explanation as to how this happened. As such, he proposed an old idea, (the same idea as Aristotle's pangenesis); explaining that the body carried 'gemmules' that would bear the qualities of one generation to the next. Two possibilities came from this proposition; one, that some qualities, such as skin color, would mix; the other that qualities such as gender would stay distinct. Various critics noted that the idea of blending qualities, as it was then known, would not uphold good qualities, but rather weaken them.

Gregor Mendel's revolutionary work could explain the principles of inheritance, but was ignored until the 20th century. Mendel and Darwin were contemporaries; while we know that Mendel read Darwin's work, we do not know whether Darwin read his. It thus begs the question: "Could Mendel's work have helped Darwin explain his theory?"

THE LIFE AND DEATH OF JOHANN MENDEL

In 1822, Anton and Rosine Mendel, two German-speaking farmers in Brunn (now Brno in the Czech Republic), were blessed with a baby boy. They named him Johann Mendel. Mendel was a sickly child, living in a poor society; he worked on a local farm a couple of days a week, but had to remain in bed most of the time. Luckily, his teacher noticed his potential and recommended him for higher education. Since Mendel was only a farmer's son, his parents could not afford it; therefore, he entered an Augustinian monastery.

In 1847, Mendel became a priest. As was the process, he had to take on a new name, Gregor. Along with this, he wanted to teach but did not pass the qualifying exams - possibly because of being unprepared. (In 1850, nervous and lacking university education, he failed one of the examinations, and the monastery was advised to send him to Vienna.)

Mendel was finally accepted to the University of Vienna in 1851 and spent two years studying plant physiology, mathematical physics and experimental physics. He began his educational career in 1854 at the German Technical Institution in

Brunn, teaching physics and natural science for fourteen years, while also fulfilling his duties as a priest at the monastery. It was during this time that he experimented profusely and discovered the principle behind the enigma of inheritance of features from parents to their offspring.

When the monastery elected him Abbot in 1868, he had to give up his scientific studies as his new duties took most of his time. Mendel's health took a turn for the worse after the Austrian government, under Bismarck, enforced a high tax rate on monasteries in 1872, which Mendel challenged. On the 6th of January 1884, Johann Gregor Mendel died at the age of 62, the cause being kidney disease.

MENDEL AT THE UNIVERSITY OF VIENNA

Mendel was most fortunate during his time at university, as Christian Doppler and Frank Ungar mentored him. Christian Doppler, the pioneer of the Doppler effect that was used in developing imaging modalities such as radar, was his physics professor. Frank Ungar, meanwhile, the man who showed that plant and animal cells were fundamentally the same, was Mendel's biology professor.

WHILE AT THE MONASTERY

When Mendel first joined the monastery, he began to work in the small garden next to the monastery wall. Today, the garden houses a stone monument to celebrate Mendel's work and the public are welcomed as visitors. The monastery is also home to a library, which, in Mendel's day, housed scientific books on subjects such as arboriculture, horticulture and botany. (The library is also open to the public.)

Plants were not Mendel's only interest, though; he also took part in beekeeping, attempting to crossbreed bees. Unfortunately, he had little success. Meteorology was another of Mendel's interests; he wrote local reports and daily records of rainfall, temperature, humidity and barometric pressures. He made records of sunspots and groundwater levels, using the monastery well to measure the height of the water.

In 1870, a tornado passed over the monastery: Mendel wrote a detailed account of the phenomenon, wherein he recorded that the spiraling motion of the tornado was clockwise, not the typical counter-clockwise rotation of the northern hemisphere.

THE BEGINNING OF MENDEL'S RESEARCH

Mendel's love of nature was the basis for his being drawn towards research; he wanted to understand how plants could gain unusual qualities. While wandering at the monastery, he found unusual variations of decorative plants; he planted them next to the usual plants, growing the generations next to one another to see if qualities were transferred between generations.

At the time, Lamarck's view that the environment was what changed the qualities of the plant was widespread. Thus, Mendel began experimenting, researching inheritance; effectively, he became the first person to discover the process of transfer of qualities from generation to generation.

He found that the plants did not change because of the environment, but they had gained vital qualities from the parentage; this was the beginning of the notion of heredity. Mendel is best known for the work he did with peas; supposedly he had a green thumb, being the son of a farmer.

MENDEL'S PEAS

A scientist must first know the appropriate form of the question they wish to study; thus, Mendel had the best mindset for scientific analysis. Plant hybridization had been practiced for quite some time. Mendel wanted to find out how hybrids occurred, when made by cross breeding the same species.

Between 1856 and 1863, Mendel studied the hybridization of peas, first testing thirty-four (34) of their varieties for the constancy of traits, and selecting twenty-two (22) of them. He always examined a large number of plants to eliminate "chance effects" and thoroughly planned his experiments. Tolerantly, he experimented on a minimum of 28,000 pea plants, breeding two generations so as to study the reliability of specific qualities in self-pollination; choosing the pea plant because it was easy to control the pollination, he also wrapped the plants separately to ensure that insects did not pollinate them. Mendel artificially pollinated the plants and meticulously noted what happened.

He found that, when crossing a tall plant with a short plant, the result looked more like the tall parent, rather than being medium as predicted by the concept of the time, the blending theory. He named this phenomenon "Dominant Inheritance". Mendel applied the rules of statistics exactly and found an outline of the inheritance of qualities. He put the qualities in opposing pairs in his experiments; for example, though not limited to these:

1) Round as opposed to wrinkled seeds
2) White seed coat as opposed to gray seed coat
3) Inflated pod as opposed to constricted pod
4) Green unripe pod as opposed to yellow unripe pod
   5) Axial position of the flowers as opposed to terminal position of the flowers
6) Long stem as opposed to short stem

He cross-pollinated different kinds of peas and, through this, he realized the occurrence of dominance and segregation or separation of specific features. He noted that the qualities in the offspring were gained in numerical ratios from the parents; this started a series of experiments of crossbreeding thousands of pea plants to find the rules of quality sharing, summarized as:

Suppose A and a are the dominant and recessive alleles respectively (alternate forms of genes present at the same spot on the chromosome, one inherited from each parent) at a single locus (position of the gene) of a plant (gene or the unit of heredity as explained in later chapters), the phenotype (appearance) of whose seeds (say a round or angular pea) depends on the genetic composition at this locus.

The possible genetic compositions (genotypes) are thus AA, Aa, and aa, but Aa seeds have the same appearance as AA, and only the double-recessive seeds aa have a different phenotype or appearance. If "A" determined the height as tall and "a " as short, the phenotype or the appearance would be tall if the plant inherited "AA" or "Aa" but would only be short if it had "aa".

THE MENDELIAN LAWS OF HEREDITY

Mendel summarized his results into three essential ideas, now known as the *Mendelian Laws of Heredity*. His theory was that certain qualities existed because of the basic elements of heredity that were shared, which we today call genes.

Mendel's first law is known as the principle of segregation; he found that, while the procreative cells formed, the fundamental elements of inheritance would separate, meaning that a cell would hold only one of the qualities, but not all. His experiments showed that qualities do not combine but stay whole, explaining why two offspring from the same parentage can have different qualities.

Mendel's second law is known as the law of independent assortment; many qualities were inherited separately; thus long- stem plants could have round or wrinkled seeds (inherited from a short-stem parent).

Mendel's third law stated that the qualities that were inherited were the outcomes of two qualities working together, one quality from each parent.

Mendel created two ideas from what he noted: 'dominant' and 'recessive'; these words are still in use today. 'Dominant' demonstrates the quality that is evident in offspring, while 'recessive' qualities are hidden by the dominant quality but may appear in future generations when they get paired with another recessive gene (both parents contributing recessive genes for the same feature)

IN CONCLUSION

Mendel introduced his famous essay, *'Experiments in Plant Hybridization'*, with the words:

> *"Courage it is indeed required to undertake such extensive labor. But it would seem to constitute the only right way of ultimately achieving this solution to a question which is of inestimable importance in connection with the evolutionary history of organic forms."*

Mendel took 20 years of work and laid it out in four short papers; he did not want attention, he wrote humbly. He only wanted to present the laws to the famous botanists of that time, who, unfortunately, did not approve of his findings. His work was not recognized until 35 years after its publication, although it opened to us the understanding of heredity. He became known as one of the best researchers of the scientific world after his work was rediscovered and accepted.

Today, Mendel is known as 'The Father of the New Field of Genetics'. The rediscovery of his remarkable experiment started the relentless scientific dash to isolate and characterize the magic molecule, of inheritance we call DNA.

SOURCES & REFERENCES – CHAPTER 14

Edelson, Edward. *Gregor Mendel: And the Roots of Genetics*. Oxford University Press, 2001.

CHAPTER 15:
LORD OF THE FLIES: FRUIT FLIES FOLLOW THE LORD'S COMMAND

"Fruit flies offer a cheap, fast pipeline to reach understanding of complex biological questions which then can be translated into medical applications."

- Professor Andreas Prokop

In 1907, at Columbia University, a tiny 16x23 feet room on the sixth floor of the Department of Zoology became known as the 'Fly Room.' It was used to breed fruit flies; it had eight desks full of incubators, milk bottles full of flies and bananas – used to gather the flies, as fruit flies eat rotting fruit.

THOMAS HUNT MORGAN

The year Gregor Mendel released his famed work was 1866. That same year, on September 25th, Thomas Hunt Morgan was born to an illustrious southern Kentucky family. Morgan's great-grandfather was Francis Scott Key – the man who wrote the 'Star Spangled Banner' – and his father was a Confederate officer. From early childhood, Morgan was drawn to biology, gathering birds' eggs and small fossils.

He was an exceptionally intellectual child. At the age of sixteen, Morgan already enrolled in the state college of Kentucky (present day University of Kentucky), and graduated in 1886 as Valedictorian. At the age of twenty-four, Morgan had a PhD from Johns Hopkins University, having done his thesis on the development of sea spiders. After he received his PhD, he was offered a fellowship to go to Europe, where he was given the chance to work for the Marine Zoological Laboratory in Naples. From this experience and contact with other scientists during his revisits to Naples, he decided to venture into the realm of experimental embryology; later, these same scientists who influenced Morgan collaborated with him on his work.

Morgan began teaching at Bryn Mawr College, where he met Lillian Samson – a graduate student in biology. They married on the 4th of June 1904, and had four children. At first, Lillian gave up her scientific career to take care of their children, but she had great influence on Morgan's work with fruit flies.

While at Bryn Mawr, Morgan had to split his time in between teaching and research. He was then asked, by his mentor and friend, EB Wilson, to go to Columbia University to become a professor in experimental zoology. Columbia University gave him the chance to pay more attention to his experiments and research, and through Wilson's insistence, Morgan changed his field of study from embryology to the mechanisms of heredity and evolution.

Morgan had doubts about Gregor Mendel's theories. He acknowledged biological evolution - thanks to the evidence thereof – but did not acknowledge Darwin's explanation of gemmules. Hugo de Vries, a Dutch biologist, had suggested that new species formed because of mutations, which is the formation of a new genetic material (gene). In the year 1900, Morgan started exploring the phenomenon of mutations and its effect on the inheritance of characteristics, with his fruit fly experiments.

In the end, his work strengthened Mendel's theory and he thus converted to Mendelism.

THE FLY AND THE LITTLE ROOM

Thousands of fruit flies can be kept in a 1-quart bottle, as they are approximately 3 mm in length, and they are also very easy to breed. The flies reproduce throughout the year, resulting in a new generation of flies every 12 days. The females are easy to differentiate from the males as the embryo forms outside of their bodies; this also makes it very easy to study by use of a magnifying glass or microscope. The fruit fly has only four (4) pairs of chromosomes, whereas humans have a total of twenty three (23) pairs.

Morgan first wanted to experiment on rats or mice, but found that they bred too slowly and were not useful when studying heredity. A better choice to study was the fruit fly.

The little room, filled with roaches, hanging bananas and pieces of scrawled note papers, gave way to several of the greatest ideas of the decade. The basis for the chromosome theory of genetics had its roots here. It was in this room that the pathways of inheritance, passing of traits from one generation to the next and their relationship to each other were identified and recorded. They proved De Vries' concept of mutations producing genetic changes. Mutations can be favorable, which allow natural selection to propagate them, or harmful, which generally will eventually be eliminated in the competition for survival.

Sturtevant, one of Morgan's students said that the room always beamed excitement and enthusiasm:

> *"There was a person at each desk working on their own experiments but each knew exactly what the others were doing. Each new result was freely discussed, there was little attention paid to priority ... or to the source of new ideas or new interpretations. What mattered was to get ahead with the work. There was much to be done. There were many new ideas to be tested and many new experimental techniques to be developed."*

Morgan's personality created this atmosphere. He had great enthusiasm, along with a serious intellect, was generous and open-minded, combined with a sense of humor. His way of teaching differed from the standard that was applicable during the 19th century at the great American universities – Harvard, Hopkins, Columbia and Chicago – that used the German research method, wherein a scientific leader set the priorities and experiments, directing his juniors to carry them out. Morgan, however, ran his laboratory with independent values and meritocracy, instead of supremacy. His students were free to come up with ideas and new ways of experimenting. There were free-flowing discussions and sharing of observations. This eventually became the model of teaching and learning in US Universities. During this time, biology and science were also shifting from observation and assumption to investigation and testing.

CHROMOSOMES X AND Y

The idea behind Morgan's breeding was to breed one fly that had a different appearance from the rest; a fly that had experienced a mutation, changing one of the regular qualities, as de Vries had noticed in his plants, with the hope of proving de Vries' theory of mutation. However, after two years of breeding, Morgan had not found a single fly with a mutation. Then, in 1910, everything changed.

Usually, fruit flies have brick red eyes, but in the April of 1910, one male with white eyes was bred, a sole unprompted mutation. Morgan's next crucial action was to mate this white-eyed male with a pure, red-eyed female. This experiment was named F1 generation. The experiment, however, only resulted in red-eyed flies. It was only in the next generation, F2 (brother and sister mating), that there were again white-eyed offspring; three red eyes to every one white eyed.

Morgan's anticipation was that there would be an identical number of male and female flies with white eyes, but what he found was that only the male flies were born with white eyes, and the females all had red eyes.

In 1910, scientists already knew that chromosomes were found in sets of two, which means that fruit flies had four pairs of chromosomes. These fine, thread-like structures were found in the center of the nucleus of cells, seen under a microscope, but we did not know their function. Morgan noticed that the four pairs of fruit fly chromosomes were all very similar, except the male fly had one pair that had two different chromosomes, while, in the female, they all looked the same. Morgan found that the male fly received chromosomes from both parents, one from the mother called X and the other from the father named Y. As a result, Morgan could use this as evidence as to why only the males were born with white eyes, as the gene that influenced the eye color was located on the Y chromosome. Morgan defined the chromosomes in this way:

> *"The egg of every species of animal or plant carries a definite number of bodies called chromosomes. The sperm carries the same number as the egg. Consequently, when the sperm unites with the egg, the fertilized egg will contain double the number of chromosomes. For, each chromosome contributed by the sperm has a corresponding chromosome contributed by the egg; i.e., there are two chromosomes of each kind, which constitute a pair."*

THE FLY PEOPLE

The first recorded fruit fly mutation was the white-eyed fly of the Fly Room, found by Thomas Hunt Morgan in 1910. But Frank Lutz of the Carnegie Institution told his peers that his flies were doing 'tricks with their wing veins', the veins that form patterns, from mutations. Since then, more and more mutations have been found, creating an entirely new area in biology.

There have been flies with strange eye colors, such as pink, purple or maroon, as well as flies with shortened or minute wings, or having no wings at all. Records have been made of flies with a large amount of hair and flies with no hair to speak of; flies with huge embryos or legs in the place of their feelers or mouths. Flies have been found with working eyes on their wings, legs or feelers; flies with two sets of wings or flies without a head.

At the end of the 1980's, the fly people – the name given to the fly researchers - the 'Drosophilists' from the Latin name for fly, Drosophila melanogaster – had noted around 3,000 mutants of the flies, giving us a large amount of biological information to work with.

> *"The way you find out what a gene does is by generating a mutation and looking at the consequences—looking at what the fly does when that gene loses its function. If you had no idea what a car engine was doing or how it worked, you'd take out different parts to see what happens. This is the basic logic geneticists use to see what function genes have."* – Herman Steller

This is what Hermann Steller, a scientist and neurobiologist at The Rockefeller University, has to say. According to the fly people, the number of mutations has, in 100 years, educated biologists in biological processes, functions of genes, and patterns of inheritance of traits more than any other animal. (Perhaps now the roundworm, *C.elegans*, may be replacing it, depending on whom you ask.)

The fly people can now, almost precisely, explain how a fruit fly embryo is made; which genes are active during development and what the genes do, as well as why they do it. Morgan's work aided the start of our understanding of biology and the basic genetic tools for growth. Stanford University's Matthew Scott believes that, if evolution finds a single useful tool, then it is a tool it will use over and over.

IN CONCLUSION

Gerald Rubin, a professor of genetics at the University of California, has worked on the *D. melanogaster* for 25 years, and said the following:

> *"If you just have the sequence of the human genome you won't know how to interpret it. It's written in a foreign language that you can't read. What you need to interpret it are the genomes of model organisms, in which people have already done a lot of work to determine what the functions of the genes are. [...] You can't do the kind of experiments in humans that you can do in the model organisms. You can't say, 'I want to cross that person with that person and see what their grandchildren are like.' So you have to go back and look in these other genomes for clues to how the human genome works."* – Gerald Rubin

From the mid 1970's, scientists have been replicating the genes of fruit flies, one by one, since recombinant DNA technology first appeared. Yet, fruit flies remain intricate creatures, despite the fact that they appear simpler than humans.

> *"The genes interact with each other in very complicated pathways and networks. We've tended to oversimplify biology to fit what we could work on in the laboratory. We've had some successes. People have figured out*

development in the early Drosophila embryo, for instance, but that's probably about the most complex thing that we have figured out with this one-gene-at-a-time kind of approach." - Gerald Rubin

Thomas Hunt Morgan's hard work was rewarded with a Nobel Prize in 1933. He was also the first American to be awarded a Nobel Prize in Physiology or Medicine.

SOURCES & REFERENCES – CHAPTER 15

Shine, Ian, and Sylvia Wrobel. *Thomas Hunt Morgan: Pioneer of Genetics.* University Press of Kentucky, 2015.

Brookes, Martin. *Fly: the unsung hero of twentieth-century science.* Harper Collins, 2002.

CHAPTER 16:
NOBEL INJUSTICE: AVERY ESTABLISHES DNA AS THE CARRIER OF HEREDITY

"Don't worry when you are not recognized, but strive to be worthy of recognition."

- Abraham Lincoln

In 1918, pneumonia overtook tuberculosis as the biggest killer in humans. However, it wasn't the first time in history. Hippocrates (3rd century BC) and Maimonides (12th century AD) described symptoms of this deadly disease with accuracy as it occurred during their lifetime.

The common pneumonia-causing bacteria named *Streptococcus* was officially identified for the first time in 1882. But, despite being discovered more than a century ago, pneumonia remains the leading cause of death in the world, with more than 1.4 million deaths annually.

Oswald T. Avery's work on pneumonia set the wheel in motion for discoveries that proved that DNA was the gene carrier; the mysterious molecule which is responsible for transferring traits from one generation to the next.

EARLY LIFE – AVERY WAS DESTINED TO BECOME A CLERGYMAN BUT ENDED UP A PHYSICIAN

Oswald T. Avery was born to Joseph Avery, a Baptist minister from England. Oswald's father moved to Halifax, Canada in 1873, where he served as a pastor. After spending nearly fifteen years in Canada, he moved to New York City, where he served as the pastor of *Mariners Temple Baptist Mission Church.* Oswald Avery was his second of three sons.

Avery's father and his elder brother, Ernest, had died by the time Avery was fifteen. While he continued his education, he had also to take on the paternal role for his younger brother. His hard work paid off when he earned his BA degree from Colgate University, where he excelled in Greek and literature. To everyone around him, it was obvious Avery would follow in his father's footsteps as a clergyman. Instead, Avery chose a career in medicine and attended the *College of Physicians & Surgeons* in New York. Here, he earned his medical degree in 1904.

After obtaining his degree in medicine, Avery soon became frustrated with the situation in medical practice, as few serious diseases could be cured. This was the era before antibiotics were discovered. He decided to pursue research and joined Hoagland Laboratory; the first privately endowed bacterial research facility in the US. At the lab, Avery decided to investigate the biological activities and chemical composition of pathogenic bacteria. He applied a systematic and meticulous approach to understanding the nature of bacteria from a study of their chemical composition.

AVERY AT ROCKEFELLER INSTITUTE AND THE CURE FOR PNEUMONIA

Recognizing the importance of his work, the *Rockefeller Institute of Medical Research* hired Avery to develop medication for pneumonia. Since antibiotics were unknown at that time, Avery was asked to work on a serum. Since there were various known types of the pneumococcal bacteria, creating a serum proved a complex process. To combat this problem, Avery developed an efficient method to quickly determine the type of pneumococcus. With the proper dosage, Avery and his colleagues succeeded in developing a serum that disturbed the bacteria's ability to ferment carbohydrates, making it impossible for the bacteria to grow in a culture.

WORLD WAR I AND AVERY'S RESEARCH ON INFLUENZA

When the United States officially entered the First World War, Avery eagerly joined the U.S. Army and served as a private, due to his status as an alien. However, his service in the army helped him earn U.S. citizenship and propelled him to the rank of Captain before the end of the war. It was during his Army service that he had the opportunity to research global epidemics, such as influenza. His contributions eventually helped to fight the Spanish flu pandemic that killed somewhere between 50 and 100 million people worldwide from 1918 to 1920. Avery, now a U.S. citizen, returned to civilian life. He continued his research in a bid to find a cure for pneumonia.

AVERY DISCOVERS A NEW METHOD TO IDENTIFY THE TYPE OF PNEUMOCOCCUS BACTERIA

There are several types of pneumococcal bacteria. Some cause infections while others are harmless. Avery, along with his colleague, Dochez, was eventually able to separate 'soluble substances' found in pneumococcal bacteria, which were different in each type. Since these substances could be recovered from a patient's urine, it ended the need to perform complex tests to identify the type of pneumococcus bacteria. Avery named this substance SSS.

THE DISCOVERY OF 'SUGARCOATED MICROBES'

Although Avery was able to separate SSS from the bacteria successfully, he didn't understand its chemical nature. He recruited Michael Heidelberger to assist him in his research and to help resolve the mystery of these mysterious 'soluble substances'. The two eventually figured out the mystery, as they learned that SSS was derived from the external bacterial capsule that was also responsible for the immune response in patients. Since its chemical composition was similar to that of sugar, Avery called it the 'sugarcoated microbe.'

To further look into the true nature of this capsule and to study its organic structure, Avery and Heidelberger recruited Walther Goebel who was a famous organic chemist at that time. Working together, they eventually synthesized antigens in the lab, that could be used to make a serum that offered better protection against certain types of pneumococcus. Moreover, they learned that the capsule in pneumococcus bacteria was the reason it was causing infections in humans. The bacteria without this capsule were harmless and didn't cause any health issues. This finding led Avery to conclude that the best way to fight these specific bacteria would be to remove the capsule.

AVERY AND RENE DUBOS – THE DISCOVERY OF THE FUNDAMENTAL UNIT OF HEREDITY

In 1927, Avery met a young graduate student named Rene Dubos at a dinner. Rene was researching microbial decomposition of cellulose in the soil at the New Jersey Agricultural Experiment Station. Avery, impressed by Dubos' work and finding similarities with his own research, offered young Rene a fellowship at the Rockefeller Hospital.

While working with Avery, Dubos was able to isolate a bacterial enzyme that could break down the capsule of Type III pneumococcus. They called this special enzyme S111. S111 played a significant role in the understanding of DNA, the molecule that eventually turned out to be the factor that transfers heredity.

FREDERICK GRIFFITH'S DISCOVERY OF STRAINS

While Avery was researching the bacteria responsible for pneumonia, Frederick Griffith also worked on the same microbe while serving in the British Ministry of Health. He discovered that the pneumococcal bacteria had two distinct forms. One form appeared as a shiny (S) object under the microscope with a smooth surface, while the other looked rough (R) and irregular. The shiny strain caused disease, while the rougher bacteria did not induce illness and were harmless. These findings corroborated Avery's discovery of the 'sugarcoated microbe'.

THE 'TRANSFORMING PRINCIPLE' – AVERY REJECTS CONCLUSIONS

In his landmark experiment in 1928, Griffith found that these bacteria could change their form from S to R in a lab, or inside a host's body. He injected mice with the R form of type I pneumococcus (harmless), and mixed it with heat-killed S-form type II pneumococcus (disease-causing, but dead). The result? The mice died from pneumonia. The samples taken from mice's blood indicated the presence of living S-type II pneumococcus. It was a remarkable discovery that showed that R-type, non-virulent bacteria 'transformed' to S-type virulent bacteria inside the body of the patient, by taking up some 'transforming principle' from heat-killed S bacteria. Another obvious conclusion from this experiment was that the R form converted to the S form. It was biologically not possible, as it would mean the transformation of one type to another. Avery rejected this conclusion, blaming lack of proper control during the experiment for these strange results.

Yet, it was the first experiment to suggest that bacteria could transfer genetic information through a process later known as 'transformation'. Future experiments isolated a 'material' that transferred the genetic information and called it DNA.

Many years later, Griffith's research was supported by conclusive evidence.

Some of Griffith's experiments were replicated in Germany, while scientists at the Rockefeller Institute reached the same conclusion after conducting a series of experiments of their own. A scientist succeeded in removing the outer covering of the S form, the virulent bacteria. Other experiments indicated the presence of a 'transforming principle' inside the cell that was responsible for changing one type of pneumococcus bacteria into another.

AVERY'S CONTRIBUTION TO THE 'TRANSFORMING PRINCIPLE' AND HIS DISCOVERY OF DNA AS THE GENE CARRIER

With data coming from various scientific sources, Avery slowly changed his mind and accepted the newly established 'transforming principle'. A few years later, he fell ill, only to return to work five years later to continue his research to gain a better understanding of the underlying biological mechanism behind the 'transforming principle'. He took a step-by-step approach, and worked on improving the accuracy and consistency of his experiments so that they were easier to re-create. He meticulously purified different components of harmful pneumococcal bacteria and injected them into the harmless type. These harmless ones, with a single injected component from the harmful type, were introduced into mice. The Eureka moment came when purified deoxyribonucleic acid (DNA) from disease-causing bacteria was injected into harmless types; they suddenly became capable of killing the mice. The other chemical components like proteins, carbohydrates, etc., when taken from harmful ones and injected into harmless bacteria, did not change them

Avery's experiments brought about better understanding as to the true nature of the transforming principle, which he proved to be deoxyribonucleic acid. He used highly purified deoxyribonucleic acid to change the R form of pneumococcus into stable S form. These experiments showed that DNA was the reason transformation occurred in pneumococcus bacteria. At the time Avery made this important discovery, DNA was considered too simple a molecule to carry such an important 'material', i.e., genes. Instead, proteins, which were complex molecules, were thought to be a potential candidate for transferring genetic characters.

Based on their careful experiments to identify the true nature of the 'transforming principle', Avery and his colleague McCarty, concluded that the transforming principle was DNA. In his own words:

> *"It means that nucleic acids are not merely structurally important but functionally active substances in determining the biochemical activities and the specific characters of cells and that by means of a known chemical substance it is possible to induce predictable and hereditary changes in cells. This is something that has long been the dreams of geneticists."*

Today, Avery's paper (1944) is regarded as one of the most important scientific works of the 20th century in the field of biology. Despite stern opposition from some scientists, especially from his Rockefeller colleague, Alfred Mirsky, Avery insisted that his conclusions were based on solid experimental work. His work was not immediately accepted. He was nominated for the Nobel prize several times, beginning in the early 1930's. However, DNA and nucleic acid research were not at the forefront of scientific activity at the time; few scientists were engaged in studying them. His work was not accepted universally, as people thought his results were from contamination. He presented further details of his experiments to allow the skeptics to understand the transforming principle and the true composition of DNA. Avery died in 1955. The scientific community took a long time to recognize and accept his seminal work, but the world realized, after his demise, he had been an outstanding giant in the field of science.

Eventually, a leading biochemist of the time, Erwin Chargaff, decided to research nucleic acids in depth, and made various important discoveries that indicated the presence of varied DNA structures in different species. Later work showed that DNA was responsible for transmission of traits from one generation to the next. In 1953, years after Avery's important work, Watson and Crick published their research in which they provided the correct model of DNA.

IN CONCLUSION

The story of DNA started with Johann Frederich Miescher, studying white blood cells and identifying a new type of enigmatic substance (DNA). Oswald Avery's work defined the function of this molecule. However, the scientific community did not quickly accept Avery's work. Unfortunately, by the time it was duplicated and corroborated with other discoveries, firmly establishing DNA as the genetic material, Avery had passed away. His work did not earn him the coveted Nobel Prize, as Alfred Nobel's will did not permit posthumous awards.

SOURCES & REFERENCES: CHAPTER 16

The Oswald T Avery Collection. *Profiles in Science*: US National Library of Medicine. [Online] Available at: https://profiles.nlm.nih.gov/CC/

CHAPTER 17:
IMAGINATION OVER EXPERIMENTATION; WATSON AND CRICK - THE DNA MODEL

"Creativity is putting your imagination to work, and it's produced the most extraordinary results in human culture."

- Ken Robinson

"My dear Michael,
Jim Watson and I have probably made a most important discovery. We have built a model for the structure of des-oxy ribose nucleic acid (read it carefully) called DNA for short." – Francis Crick

The above is an excerpt from the handwritten letter Nobel-prize winner, Francis Crick wrote to his 12-year-old son, Michael, on March 19, 1953. The letter was sold for $5.3 million ((£3.45 million) in 2013 at a New York auction.

Although many people believe that Francis Crick and James Watson discovered DNA, it's simply not true. They came up with a model to describe its chemical structure. The American biologist and English physicist developed their ingenious model based on research that other scientists had done. Watson and Crick's DNA model in the 1950's, actually used all the available information and turned it into a perfect model for DNA. Their brilliant work helped scientists understand DNA's structure and various processes involving the molecule.

SCIENTISTS WHO CONTRIBUTED TO WATSON AND CRICK'S DNA MODEL – THE GROUNDWORK

FRIEDRICH MIESCHER

Friedrich Miescher was born in 1844 in Switzerland. From childhood, he had a passion for learning, despite the fact that he had a hearing impairment. When he graduated, he did not practice medicines but chose a career in medical research. As a research associate, Miecher joined *Hoppe-Seyler Laboratory* in Germany, which was the first lab in Germany focused on tissue chemistry. Scientists at Hoppe-Seyler were already isolating and studying molecules that provided the building material for cells.

Miecher focused his research on white blood cells. While studying these cells, he found a new chemical in their nucleus and termed it 'Nuclein.' Made of oxygen, hydrogen, nitrogen, and phosphorus, the mysterious molecule was also found in other cells. These remarkable discoveries took Miecher only a year after he joined the lab. Unfortunately, his work wasn't published until 1871, after the laboratory confirmed the results. Miecher discovered Nuclein and analyzed it years before other scientists. Yet, throughout his life, Miecher believed that proteins were the molecules that carried traits from one cell to the next, and recognized them as molecules of heredity.

PHOEBUS LEVENE

Phoebus Levene, born in Russia, pioneered the study of nucleic acids. He moved to the U.S. after graduating from St. Petersburg Imperial Medical Academy. Later, Levene enrolled at Columbia University to continue his studies.

During his research, he met and worked with well-known chemists. Among those were German chemists, Albrecht Kossel and Emil Fischer, who were studying proteins and nucleic acid. Their research inspired Levene, and he, from then onwards, devoted his life to research. Recognizing his talents and dedication to research work, Levene was appointed head of the biochemical laboratory at the *Rockefeller Institute of Medical Research* in New York. During his time at the institute, Levene worked mostly on nucleic acids.

While investigating different types of nucleic acids, as they were known at that time, Levene discovered that the nucleic acid found in different organisms had a different composition. When he studied the acid found in the thymus of animals, he noted that it contained four nitrogen compounds, including thymine. Surprisingly, thymine was missing when he investigated the nucleic acid found in yeast. Scientists during Levene's time knew that nucleic acids also contained carbohydrates and phosphorus. However, the function and structure of these compounds were not known. Kossel continued his research to learn more about these mysterious ingredients and learned that carbohydrates found in the nucleic acid of yeast were actually a five-carbon sugar. He named it 'ribose'. He also discovered that nucleic acids found in the thymus contained a different five-carbon sugar. He called it deoxy-ribose.

Levene and other scientists of his time analyzed DNA and broke it down into smaller chemical components. Levene showed that a nitrogen base, a pentose sugar, and a phosphate created a unit inside DNA. He called this unit the nucleotide. According to his work, DNA could consist of strings of nucleotides. He also thought that each DNA molecule consisted of only four nucleotides (known as the 'tetranucleotide' hypothesis), a claim that was proved wrong by later research.

Levene did not believe that DNA carried the genetic code. He thought the molecule was too simple to have such an important function. He, like others at that time, believed that proteins were the only molecules with a level of complexity that could allow them to function as gene carriers. It was only a few years after his death that scientists started to accept that DNA was a gene carrier.

ERWIN CHARGAFF

Erwin Chargaff received his doctoral degree from the University of Vienna and left his homeland to pursue a fellowship at Yale University in the United States. He worked on TB bacteria for a few years, then moved to Europe, but returned to the US to join Columbia University as a professor, where he continued to work for the rest of his career. When Chargaff came across Oswald Avery's report on DNA as a hereditary unit, he was excited. In his own words:

> *"Avery gave us the first text of a new language, or rather he showed us where to look for it. I resolved to search for this text. Consequently, I decided to relinquish all that we have been working on, to bring it to a quick conclusion."*

It was the moment of inspiration. Chargaff decided to shift his focus to study nucleic acid. He thought DNA for different species would have different structures and characteristics, as every species had different hereditary traits. He stressed that scientists should be able to show the variation between different DNA using the latest technology, especially chromatography.

Chargaff himself used available technologies and skillfully analyzed the DNA of different species. He sent his findings for publications, but they were surprisingly returned with objections that sounded unreasonable, even at that time, especially in light of the available research data on DNA. Eventually, his paper was published, followed by a review of his findings in 1950. This review became known as Chargaff's rule. The paper showcased two major findings:

1. DNA is made from a backbone of four nitrogen-based chemicals called nucleobases, named Adenine, Guanine, Thymine and Cytosine.
2. For every DNA sample, the number of guanine (G) units matches the number of cytosine (C) units and, similarly, the number of adenine (A) units and the number of thymine (T) units is always equal.

LINUS PAULING

From his time as a young student, Linus Pauling, was fascinated by chemistry. He actually set up his own laboratory in the basement in order to conduct chemistry experiments.

Linus' father died when he was only nine years old, creating financial hardship, which forced Linus to work his way through college, to pay for his studies as well as living expenses. At the early age of fifteen, Linus entered Oregon State University, after deciding to leave school without completing his diploma. The school refused Linus the flexibility to complete credit hours for American history while attending college at the same time. However, he was later awarded a diploma when he became the world's top scientist and received two Nobel prizes. During his time at college, scientists became interested in exploring the electronic structure of atoms. Their aim was to understand how atoms bonded to form molecules. Linus found it interesting as well, and decided to focus his research on the relationship between the physical and chemical properties of substances, based on their atomic structure.

Linus joined Cal Tech in 1922, where he completed his Ph.D. in physical chemistry and mathematical physics. During his graduate research work, he focused on the X-ray diffraction technique and used it to understand the structure of crystals. From his studies, Linus proposed a triple helix structure of DNA, which proved to be wrong later. However, he explained the structure of protein molecules and the nature of chemical bonds correctly. Today, Linus is known as one of the greatest scientists of the 20th century.

JAMES WATSON AND THE DISCOVERY OF DNA STRUCTURE

James Watson was born in Chicago. He used to go to the library with his father every Friday night from a very early age. At fifteen, Watson was awarded a scholarship to attend the University of Chicago, where he completed his graduation in zoology. During his college days, Watson read, "What is Life," by Erwin Schrodinger. In his famous book, Schrodinger discussed the possibility of all the information of life contained in a molecule. It was a fascinating speculation that caught the imagination of James Watson and inspired him to focus his research on this mysterious molecule. During his thesis, James Watson studied the effects of X-rays on viral multiplication. During his postdoctoral work, he went a step further and studied the structure of DNA in viruses, even though he mostly found it boring. Luckily, he met Maurice Wilkins, who was doing similar work but using a different technique to take pictures of crystals. He found it very interesting and exciting. He tried to find a place in Wilkins' lab but was unsuccessful.

WATSON MEETS CRICK AT CAMBRIDGE UNIVERSITY

Watson was fortunate to obtain a position at Cambridge University to work on proteins. DNA, by now, was firmly established as the molecule which carried the material of heredity or genetics. At Cambridge, he shared an office with another graduate student, Francis Crick, who was a crystallographer. They became fast friends. Watson's interest and enthusiasm for discovering the structure of DNA won over Crick, who also decided to pursue it and they started investigating the structure of DNA together.

Watson and Crick used to discuss the possibilities of expanding the study of DNA. Watson had read Pauling's work in which he investigated the structure of protein molecules, using X-ray photographs. But unlike Pauling, Watson and Crick were not chemists. It was a barrier that they overcame by using their imagination and energy to solve the ultimate puzzle of life. They feverishly worked on building models with cardboard pieces, using all available data. They had a few lucky breaks. Several scientists, including Linus Pauling, Maurice Wilkins, and Rosalind Franklin were all working towards the same goal. They both met with Wilkins, who had even taken some unclear pictures of DNA strands by the time he met Crick and Watson. Rosalind Franklin, a female scientist who had studied science at Cambridge, joined Wilkins in 1951. Prior to working with Wilkins, she had researched coal and graphite microstructure and learned X-ray diffraction techniques. She helped Wilkins take better pictures of DNA by making use of X-ray diffraction.

THE DOUBLE HELIX MODEL OF DNA

Several scientists, including Linus Pauling, Maurice Wilkins and Rosalind Franklin, were all racing to become the first to decipher the structure of this magic molecule. Watson and Crick invited Wilkins and Franklin to examine their first model. To their disappointment, Rosalind Franklin found their work inaccurate. Disappointed, but not ready to give in, Watson gathered more data on DNA.

Without her permission, they were shown a clear picture of a DNA molecule taken by Franklin; it was the famous photograph 51, which showed that the DNA had 2 strands. It was a critical piece of information. They also saw a model by Linus Pauling, which, though not accurate as it had 3 strands, had several crucial clues; and finally, they saw the work of Chargoff, which showed that the base pairing of nucleotides was in a fixed pattern. Armed with these sets of data, they feverishly worked on building new models with cardboard pieces and sticks, using all available information; without actually doing any of the basic work, Watson and Crick ingeniously created a model which turned out to be perfect.

THE NEW AND IMPROVED MODEL

They invited Wilkins to evaluate their new model. Wilkins went back and compared the model with Rosalind's photograph and confirmed that it was a compatible model. Both labs submitted their research simultaneously, and it was published in the prestigious scientific journal, *Nature*, on 25th April 1953. In 1962, Watson, Crick and Wilkins received the prestigious Nobel Prize for their brilliant work. Rosalind Franklin was not nominated, as she had passed away in 1958.

IN CONCLUSION

Finally, by the 1950's, the mysterious phenomenon of transmission of hereditary traits from parents to children, which had puzzled scientists for centuries, was discovered. The molecule, DNA and its chemical structure were figured out.

This set the stage for a more exciting chapter in science, advancing from understanding nature to manipulating nature.

SOURCES & REFERENCES – CHAPTER 17

Olby, Robert Cecil. *The path to the double helix: the discovery of DNA.* Courier Corporation, 1994.

CHAPTER 18:
DNA SLEUTHS: CRACKING THE CODE

"Facts which at first seem improbable, even on scant explanation, drop the cloak which has hidden them and stand forth in naked and simple beauty."

- Galileo Galilei

James Watson and Francis Crick's discovery of the molecular structure of DNA had the scientific world in a roar of excitement as to the possibilities. Scientists began studying the body's design and how it worked. How does the genetic information carried in the DNA actually work? How does it make harmless bacteria which receive the DNA of disease-causing bacteria transform into a lethal species? How does the DNA from the sperm and egg of parents produce the flesh, bones and skin of a baby? The structural elements of a body are mostly made of proteins. What is the connection between DNA and proteins?

The question posed at the time was: *"How do DNA molecules create proteins that comprise the outline of the body, the enzymes that perform the chemical developments needed to live, to develop and to reproduce?"*

GAMOW

George Gamow was the author of the Mr. Tompkins science book series, which described a bank clerk with a big dream. This dream led him to meet Einstein, Watson, Crick and other leading scientists, while he took it upon himself to study the difficult scientific notions of the era. Gamow was born in the part of Russian Empire known today as Ukraine. After studying in Russia and Germany, Gamow worked at Cambridge University until he moved to the U.S. in 1934, where he worked at George Washington University in Washington D.C. Gamow is famous for his work on the radioactive decay of atoms – his work supported the Big Bang Theory. During his lifetime, Gamow tried to solve the riddle of connecting DNA to protein after the publication of DNA design fired his curiosity. His concept of a DNA code was helpful in pursuit of deciphering the mechanism by which DNA operated.

DNA AND RNA

Scientists knew that the production of proteins occurred inside the cytoplasm on ribosomes; meanwhile, DNA was only found in the nucleus. RNA is an in-between molecule, related to DNA, but so named because of the sugar in the molecule being ribose instead of deoxyribose, as it is in DNA. RNA was detected in both the nucleus and the cytoplasm. It was also known that proteins were comprised of long chains of twenty different amino acids, arranged in various sequences. Radiolabeling research had established that RNA was tied with amino acids on ribosomes.

Sydney Brenner, a scientist born in South Africa and who had worked at Cambridge University, showed that RNA was the connecting molecule transferring the information from DNA to the ribosomes, the factory churning out proteins. It appeared from the data available that there had to be code in the DNA, which was translated by the RNA to direct the production of proteins like a Morse code used by telegraphy. DNA had four nucleo-bases and there were 20 amino acids, which made up the proteins. If each base coded for an amino acid, it would only cover four; if a sequence of two bases coded for one amino acid, then it would cover eight. However, if a specific sequence of three bases coded for a specific amino acid, it would cover 24.

Crick hypothesized, based on the existing facts, that a triad code existed. What was this magic code?

THE RNA TIE CLUB

In 1954, Gamow and Watson teamed up to form a research group. They called it the RNA Tie Club with the motto: *"Do or Die; or don't try."* The goal of the members was to decipher the relationship of DNA to protein production. The club consisted of twenty members, each named for an amino acid, with four honorary members named for the nucleotide bases in nucleic acids. Most of their communication was via letters, discussing new thoughts and notions before being published. The club met twice a year; consisting of cheerful meetings, with a lot of alcohol and cigars. As with all clubs, there were some rules, with a simple emphasis on developing close friendships that eventually led to inspired notions. Members were given black, wool ties, decorated with a green and yellow DNA helix. From a total of twenty members, eight went on to become Nobel Prize winners; yet, the Nobel Prize for decoding the DNA code did not go to any member of the RNA Tie club.

KHORANA TAKES SCIENCE TO THE HOLY GRAIL

The pace of scientific work accelerated rapidly. Har Gobind Khorana was born in a small village of 100 families in India. Though poor, his father emphasized education. Their family was the only one in the village which went to school. After obtaining a Master's degree, he received a scholarship to study for a Ph.D. at the university of Liverpool in Britain. Khorana was greatly influenced by professor Vladimir Prelog, who mentored him in Zurich, where he did his postdoctoral work. He subsequently had stints at Cambridge and the University of British Columbia, where he worked with a group focused on nucleic acids.

In 1960, he moved to the University of Wisconsin. His ingenious and meticulous work at the University of Wisconsin broke new ground by artificially synthesizing RNA in the test tube, using simple chemicals and enzymes. Wow! Science appeared to be touching the holy grail of nature: the very molecules of life were now cooked up in a test tube.

NIRENBERG

In 1959, Marshall Nirenberg joined the National Institute of Health (NIH) as a biochemistry scholar. After finishing his PhD at the University of Michigan, he decided to research the instruments used in the production of proteins. The course DNA follows to produce proteins was unknown and the messenger, RNA, was not yet discovered.

Nirenberg was an inexperienced scientist who had no formal training; as such, he had the choice of working on an already recognized task or going into the new and unknown field of molecular genetics. But his lack of education and field expertise didn't hinder his spirit; he leaped at the chance to become the one to unravel the mysteries of DNA along with the top scientists in the world.

His only companion was the German postdoctoral student, Johann Mathaei, a very capable and brilliant young scientist. Meanwhile, the RNA Tie Club's members were gathering proof that RNA molecules act as envoys for DNA inside the nucleus, to the ribosomes inside the cytoplasm. Yet, at the time, Nirenberg and Mathaei were oblivious of these new findings.

NIRENBERG AND MATHAEI'S EUREKA MOMENT

E.coli is a single cell bacterium, which is generally located in the bowels of the human body; this is the cell Nirenberg and Mathaei chose for their studies. They broke up E. coli's cell walls by using a blender, resulting in the release of ribosomes. Ribosomes are structures which produce proteins when RNA is added. They divided the solution containing blenderized E. coli into twenty different vials, each of which had a mixture of the twenty amino acids. The amino acids were then radioactively tagged, but only one amino acid was tagged in each vial - a different one in each vial. After a time in the incubator, the solution was put through a sieve to separate the unbound amino acids and RNA, which would pass through the sieve, but the ribosomes would be left behind. The ribosomes were then inspected; if one were found to be radioactive, it would mean that an amino acid had been used by the ribosome to produce a protein.

The notion here was that, if a new protein was created, one could track the amino acid used by the ribosomes, because of the radiolabeling it had undergone. After this, the two scientists added artificial RNA, created by Khorana's process, as described earlier. The artificially created RNA, with a defined nucleotide sequence, was added to the solution, creating a magic breakthrough for molecular biology research:

"RNA is made of 4 nucleotide bases, adenine (A), cytosine (C), guanine (G) and uracil (U). As these RNA chains were synthesized they could make any sequence of them like UUU or UAC etc. They theorized that if it is a triplet code, then one of the sequences like UUU should be a code for a particular amino acid. When synthetic RNA containing UUU was added to the 20 tubes containing different amino acids, one of the tubes with the right radio labeled amino acid will be transformed into a protein. If the experiment succeeded, then UUU will be the code for that particular amino acid."

- This was the famous poly-u experiment.

On 27 May 1961, at 3 am, Mathaei added artificial RNA, with a link of uracil nucleotides only, UUU, to the twenty vials. The exercise could have failed, since messenger RNA has a unique code at the start, which signals the ribosomes to initiate protein synthesis. What Mathaei added did not have this code; but on this specific day it did not matter. The solution that Mathaei was using had double the amount of magnesium in it compared to the normal solution. At the time, they did not know it, but the extra magnesium overrides the need of the RNA signaling to start protein synthesis.

Mathaei noticed something strange happening in the vials; lo and behold, there was a protein molecule located in the vial; the protein consisted of a long chain of a single amino acid, phenylalanine. The multiple amino acid chain containing only phenylalanine, thus, had to be coded by a triad code UUU, as that was the artificially created RNA that had been added to that vial. This was their Eureka moment.

Nirenberg, at the time only thirty-four years old, was not yet known to the scientific world of RNA and DNA. But in August 1961, Nirenberg presented his findings at the International Congress of Biochemistry in Moscow. Unfortunately, since he was not a renowned scientist, the presentation was held in an almost empty room. Luckily, Nirenberg had met Watson the day before, to whom he explained his findings. Watson, though he doubted Nirenberg's findings, had an associate attend the lecture, nonetheless. Watson's associate, on the other hand, was truly swayed, upon which Watson arranged for Nirenberg to present a repeat of the same lecture. This time, more than one thousand scientists attended it. The scientists were in awe, and excited by Nirenberg's findings.

NEWS REPORT

President Kennedy had announced the notion of the 'New Frontier' the year before, and thereafter this notion became U.S. vernacular, an adequate fit for the discovery. The New York Times wrote: *"The science of biology has reached a new frontier [...] it could lead to a revolution far greater in its potential than the atomic or hydrogen bomb."* The Chicago Sun Times stated that this discovery was *proof of universality of all life*. John Pfeiffer, a journalist, wrote that this discovery was of larger importance than the Russian cosmonauts that had circled the earth. Thus, Nirenberg not only rocked the world of the scientific community, but the entire population.

THE RACE FOR THE CODE

Nirenberg and Mathaei only found the code for phenylalanine; the other codes had yet to be discovered. Thus began the race to find the remaining codes.

After Mathaei finished his post-doctorate and returned to Germany, Nirenberg was left alone. NIH assisted Nirenberg because they thought that he could be the first U.S. Government employee to win a Nobel prize. The "coding race," as this was dubbed, led to a photo-finish between the National Institute of Health and New York University in 1961 and 1962; but Nirenberg finished the work and recognized the triad codes of all twenty amino acids. The Rosetta stone moment for biology had happened; the language of DNA had been decoded.

IN CONCLUSION

One last piece of the protein synthesis puzzle was left to Robert Holley, at Cornell University, to answer. He was the one who recognized 'transfer RNA'; this molecule conveyed the genetic code from DNA to ribosomes. These three scientists, Nirenberg, Khorana and Holley, were honored in 1968 with the Nobel Prize.

SOURCES & REFERENCES – CHAPTER 18

Cobb, Matthew. *Life's greatest secret: the race to crack the genetic code.* Hachette UK, 2015.

Portugal, Franklin H. *The Least Likely Man: Marshall Nirenberg and the Discovery of the Genetic Code.* MIT Press, 2015.

CHAPTER 19:
DNA GOES TO WORK: ONE GENE, ONE PROTEIN

"The vision to see, the faith to believe, and the will to do will take you anywhere you want to go."

- *Unknown*

In 1584, an unusual illness was described in a child whose urine turned black after it was released from the body. In 1866, the famous German pathologist, Virchow named the illness Ochronosis. The illness caused the pigmentation of cartilage, ligaments, tendons and blood vessels, besides turning the urine black. The malady, interestingly, was also noted in an Egyptian mummy, a newborn baby dating back to 1500 B.C. The cause of the discoloration was only discovered in 1891.

ARCHIBALD GARROD

Garrod's father, a renowned doctor and professor of medicine, did the first chemical test ever carried out in clinical medicine. The test resulted in the discovery that people who suffer from gout have more uric acid in their urine and blood. His son, Archibald Garrod was born in London in 1857; he followed in his father's footsteps in studying medicine. From early on, Garrod was fascinated by chemical pathology.

During his practice, he was assigned to the Children's Hospital. Here, alongside a well-known biochemist, he decided to take on the cases of children who were passing black urine. At the time, the general understanding was that black urine was caused by a contagious gut illness through bacteria, which in turn caused urine to turn black.

Garrod found this assumption interesting, so, in 1889, set about studying thirty-one cases of children passing black urine. After careful research, Garrod could not find any differences between the affected urine and normal urine. From his research, he thus proposed a new theory that the discoloration of urine was caused by a problem in the digestive system.

One day, William Bateson, a renowned botanist and friend of Garrod, told him to look into hereditary elements. Bateson had translated Mendel's work to English, so had some knowledge as to heredity. After encouragement from Bateson, Garrod collected detailed information from the families of the affected children, inquiring as to their medical past, and found that the families had long histories of consanguinity. He noticed a pattern of inheritance in the families, and proposed the concept of *"in born errors of metabolism"*. He thought that genes had a biochemical role, as a defect in a gene led to a biochemical change in the body, producing black urine.

Garrod shared his idea of inherited diseases with the Royal College of Doctors in 1908. His detailed study of the information and findings were revolutionary; however, those who reviewed the paper he wrote for the British Medical Journal were not impressed. The significance of his work was only accepted later as a milestone in medicine and genetics.

BEADLE AND TATUM

George Beadle was born to a farmer in Nebraska. As he grew older, his father believed that he would one day also become a farmer. But Beadle had a different vision for his future, and decided to further his academic study - especially after some motivation from his science teacher. Genetics piqued Beadle's interest while he was at college and, in 1931, he received his PhD.

Beadle then went on to Caltech, joining Morgan's research team to study the genes of the fruit fly. Here, he became increasingly interested in decoding the mutations that changed the color of the fruit flies' eyes. It was obvious that transformations of the genes resulted in the different eye colors, but the question was how it happened. Beadle came up with the theory that the genes created an element that was needed for the eye color, and the color was changed by the mutation in the gene.

Some timer later, Beadle changed positions to work on his theory at Stanford University where he met and teamed up with Edward Tatum. Tatum was born in Colorado and received his degrees from the University of Chicago and Wisconsin. After finishing his postdoctoral work, he began his career at Stanford University. Tatum was knowledgeable about nourishment, biochemistry and the genetics of microbes. Together, Tatum and Beadle separated the pigment that created the eye color of fruit flies; however, they found the work rather dull, so they started experimenting on a less complex organism, known as Neurospora, the bread mold.

NEUROSPORA

Neurospora grew fast and effortlessly in a soup in their laboratory. The soup's elements could be further manipulated to their specifications, so, one by one, Tatum and Beadle removed the elements from the soup and found that Neurospora required just simple sugars and salts. Biotin was the only vitamin needed in a minimal media (broth which had the minimal requirements to support the growth); no amino acids, the foundations of proteins, were necessary. The fungus could grow easily on bread, as well as the simple broth made in their laboratory, while other organisms couldn't manufacture most of the molecules needed to live, known as essential nutrients or vitamins. Humans, for instance cannot produce them, while Neurospora, on the other hand, can survive with just sugars, salts and Biotin, which meant it could manufacture the rest.

Tatum and Beadle thought that the fungus' cells could create the intricate chemicals that the organism needed to survive, grow and reproduce, through enzymes. The result of this theory was that, if an enzyme were influenced by a mutation in the gene that controlled it, the fungus would die if the needed element were not available. Tatum and Beadle blitzed the fungus with X-rays to force mutations, and, although the theory was good, there was no promise that the experiment would succeed. So they decided to test 5,000 mutants: luckily, they succeeded with the 299th. This mutated mold could no longer live in the minimal media, although it could survive in the normal soup.

Their next action was to add amino acids to one batch and vitamins to the other; only those with the vitamins survived. Tatum and Beadle found that Vitamin B6 was the needed vitamin, after they had put the vitamins in one at a time. The enzyme that became defective by the gene mutation clearly had to facilitate the production of B6. They crossed the mutated Neurospora with standard ones to show that the mutation in a single gene created a flaw in the function of a single enzyme. The experiments were meticulous and sophisticated, but through them Beadle and Tatum confirmed Garrod's theory regarding the inherited disorders of metabolism; a flaw in one gene creates a change or lack of molecules which the body needs to be healthy, resulting in an illness.

Tatum and Beadle were awarded the Nobel Prize for linking a gene to an enzyme, the work then known as the one-gene-one-enzyme model. Their work was later modified to one-gene-one-protein, since certain genes are only cyphered for proteins, not enzymes, while others are cyphered for RNA molecules.

IN CONCLUSION

Beadle summed up his work on the use of genes as leading to the creation of certain enzymes, in an influential article: *"A given enzyme will usually have its final specificity set by one and only one gene."*

Today, this is now known as the 'one gene-one enzyme' theory, which has been superseded by one-gene-one-protein following further studies, but Beadle and Tatum's work started a novel paradigm in genetics – the beginning of molecular genetics.

SOURCES & REFERENCES – CHAPTER 19

Berg, Paul, and Maxine Singer. *George Beadle, an uncommon farmer: The emergence of genetics in the 20th century.* Cold Springs Harbor Laboratory Press, 2005.

CHAPTER 20:
RECOMBINANT DNA: TAKING ARTIFICIAL BREEDING TO THE NEXT LEVEL

"What the public needs to understand is that these new technologies, especially in recombinant DNA technology, allow scientists to bypass biological boundaries altogether."

- *Jeremy Rifkin*

There are almost 300 trillion cells in the human body, so what could splicing the genes from a frog and introducing them into bacteria mean for humanity, and how did this type of experiment begin?

THE MEETING

Two men attended the same meeting in Honolulu during November 1972. As faith would have it, their paths crossed. They didn't know much of each other, even though they worked only a few miles apart. Upon meeting, they decided to have a sandwich at a Korean deli on Waikiki Beach. Stanley N. Cohen was a physician and a researcher for Stanford, while Herbert Boyer worked at the University of California, San Francisco. [Boyer and his wife, Gracie, had two Siamese cats, named Watson and Crick.]

As the man behind the counter cut their bread and meat, marrying the two to create a sandwich, Cohen and Boyer conversed over hot pastrami and corned beef. They spoke of collaboration in the molecular biology field, but as their server sliced away with a knife, the collaboration moved from opening new vistas to the cutting and splicing of genes. As such, the unimaginable idea of taking a gene out from one organism and placing it in a different species altogether or recombining genes from different species began to flower.

Boyer had delivered a speech about his work that day, and Cohen was keen on learning more after their paths crossed. In turn, Boyer was just as keen when Cohen explained his work. A realization then dawned on them; their knowledge was as attuned as two aspects of DNA molecules that match each other. Thus began the preparation for their experiments, over the last bites of their sandwiches.

Stanley Falkow was another professor who attended the meeting; he and Boyer spent hours together drinking Blue Hawaiians on the beach, while discussing the topic of the meeting. After their meal, Falkow and Boyer were enjoying another Blue Hawaiian in the foyer of the hotel, when Cohen joined. A little while later, Charlie Brinton, another attendee and a friend of Falkow, also joined the gathering. Brinton was not satisfied with the meal they had, so he asked the others to accompany him for another snack.

A walk down a dark road led them to a commercial area, where they found a New York style deli. Settling in a booth, they enjoyed themselves, marveling at the Hawaiian waiter, who knew the menu's details but could not pronounce all the components. After the second meal of the night, their conversation turned to their work. Everyone took the time to explain what field they were involved in, their own area of proficiency and the existing information available in that field. Throughout this discourse, Cohen had begun to thread a new idea: "Could DNA from two different species be combined?" While Cohen began to explain this, Boyer sprang to the conclusion that this should indeed be possible.

BACTERIA ARE SINGLE CELL CREATURES

Bacteria are not complex creatures; they do not have the same type of immune system as humans do. The human body has antibodies and white blood cells, which protect the body from germs. There are numerous viruses, such as those that cause colds, while others cause more serious diseases, such as AIDS; these viruses do not only attack humans, they also attack bacteria. Bacteria have immune systems of their own, although they are not alike to those of humans. They do not have antibodies and white blood cells; instead, nature provided them with enzymes that have the ability to split a virus into multiple pieces, practically slaughtering the virus. Restriction endonucleases, or in shorter form, restriction enzymes, slice DNA at precise points, always the same point, known as the recognition sequence. Each enzyme targets a different site. Around 900 enzymes have been found thus far. Bacteria can contain more than one enzyme; these enzymes attack certain sites that are not present in the host bacterial cell, so as to prevent the DNA of the bacteria being damaged.

From Geneva, Switzerland, came Werner Arber, who began researching the field of enzymes in bacteria. Based on his research, he theorized that enzymes present in bacteria cut a virus, thus preventing or "restricting" its growth.

In Baltimore, at the Johns Hopkins University, Hamilton Smith verified the theory and found and purified the restriction enzymes in the bacterium, Hemophilus influenza, which causes pneumonia in humans. Daniel Nathans, also from Johns Hopkins, furthered these findings by using these enzymes to split a virus, known as the simian virus, into eleven small pieces. His 1971 publication discussed and foresaw the potential of these enzymes. This discovery was the beginning of a new paradigm in biology, the dawn of "Recombinant DNA technology".

In October 1978, the Nobel assembly awarded the Nobel Prize in medicine and physiology to the three above-mentioned scientists; the restrictive enzymes were then referred to the 'chemical knives' or "molecular scissors", as they sliced genes into distinct pieces. This became valuable when researching gene functions and genetic disorders. Scientists now could slice out small segments of DNA and study it in detail. It is simpler to work with smaller pieces; in particular, they could isolate an individual gene, learn its function, how it can malfunction, etc. They could even cut one out from one organism and introduce it into another.

BACTERIA AND THE CHROMOSOMES

Bacteria can not only have their specific chromosomes, but also have additional DNA, known as plasmids; but these mostly consist of just a few genes. These are commonly present as a round bit of DNA that reproduces separately from the main set of chromosomes; they do not further the existence of the bacteria, but rather offer certain benefits to the bacteria, such as resistance to antibiotics. They are considered independent entities in the bacteria; however, they are incapable of existing outside the cell; thus, we do not consider them as organisms that are alive. Plasmids carry hereditary constituents, so, when they are presented to bacteria, these constituents become a fragment of the genome - the gene pool of the bacteria.

PAUL BERG

'*Microbe Hunters*' is a book published in 1926 by microbiologist Paul de Kruif; the book contains twelve tales explaining how microscopic beings were found. It became an immediate success because of the way de Kruif wrote it, using an exaggerated style and inserting fictitious discourses. As a result, a large percentage of the younger generation had their interests piqued, with many pursuing a science career. Paul Berg was one of these inspired young men who followed a science career after reading de Kruif's book. Born in Brooklyn, the same year as the book was published, Berg was introduced to the *Microbe Hunters* only during his junior high school years. His teachers, who cultivated his interest even more, suggested an afterschool science program to him. Berg's autobiography states:

> *"Looking back and nurtured in the curiosity and the instinct to seek solutions are perhaps the most important contributions education can make. With time many of the facts I learned were forgotten but I never lost the excitement of discovery."*

Berg went on to New York City College after high school and studied to become a chemical engineer, even though he was much more interested in the chemical events revealed in the genetic make-up of organisms. Eventually, Berg completed a degree in biochemistry, after which he went to Western Reserve University and received a PhD. By then, Berg was bent on doing research for academic purposes. He began showing curiosity toward enzymes and went to Copenhagen for a year, where he found a new enzyme that can be used in the making of nucleic acids. After this, he moved to Washington University in St. Louis, where he found a new organic module group, tangled in the fat metabolism of the human body. Berg's two years at Copenhagen and St. Louis changed his focus from those of a typical biochemist to becoming a molecular biochemist.

He was more and more involved in the researching of genes, and, in 1959, moved to Stanford, where he stayed throughout the rest of his career. There, Berg began experimenting on a virus known as SV40 - a minor virus that has a round shape and only five genes. Berg opened SV40's genome with restriction enzymes. He used a similar approach to cut the DNA of an E. coli plasmid. It was then possible for his team to link two pieces of DNA, one from the simian virus

SV40 and another from the bacterial plasmid, by using DNA ligase, an enzyme normally present in cells, which joins two fragments of DNA. Cells use this enzyme to repair breaks in DNA strands. The National Academy of Science defined the intricate process in their 1972 records. Berg's work was the first creation of a recombinant DNA molecule - viral DNA with an E. coli plasmid, joined together to produce a new combined DNA of two different entities.

COHEN AND BOYER

After meeting in Hawaii, Cohen and Boyer began planning experiments, the first of which was to join two plasmids together through slicing one open, using a molecular scissor. Cohen dubbed this the Chimera, using the legendary Greek monster, [made up of part lion, part goat and part snake], as inspiration. Berg's work, as well as findings from two others, gave the support these two needed to sustain and further their research.

Using a sort of bacteria usually found in E. coli, Cohen successfully joined a plasmid to the bacteria. The plasmid was identified as pSC101, which creates resistance against the antibiotic tetracycline when it exists in bacteria. Boyer and Cohen used Eco RI, a newly found restriction enzyme, and established that it sliced BSC 101, a round plasmid, in only one place; this caused BSC 101 to become a straight string instead of a round plasmid. They then added a new gene, which created resistance against Kanamycin, a different type of antibiotic, at the ends of the string that were left open. The attachment of a new gene to the two open ends was made possible through the use of the enzyme DNA ligase; this resulted in the straight string becoming a circular plasmid once again.

Oswald Avery had earlier proven that harmless streptococcal bacteria could be made into virulent bacteria by adding the DNA of a virulent one. A group at the University of Hawaii in 1972 created a chemical process used to add alien DNA to bacterial cells, through physical and chemical straining that upset the cell walls of the bacteria, which then permitted the alien DNA to enter the cell. This process, being readily available, worked in Cohen and Boyer's favor. The two then combined their recently merged DNA with bacteria and suspended the mix into a calcium chloride concentrate at freezing temperature, after which they quickly heated the mix and cooled it again. This treatment caused an occurrence known as 'heat shock', which in turn caused the plasmid DNA to go into the bacteria. Bacterial cell membranes consist of hundreds of minute openings that could, in fact, permit the plasmid to go into the cell, but the cell walls are charged negatively and thus prevent the DNA, which is also negatively charged, from entering. The calcium chloride concentrate causes electrostatic neutrality and the heat shock creates an ionic current that pushes the DNA through the cell walls; after which the cold once again hardens the cell wall. After this process, Cohen and Boyer refined the bacteria and found that it was now resistant to both tetracycline and kanamycin. So, while Berg created the first composite DNA molecule, Cohen and Boyer created the first living organism that had composite DNA.

Cohen then added the newly formed plasmids to E. coli; this caused the next generation of bacteria to carry the plasmid, resulting in the transmission of resistance to tetracycline and kanamycin. This experiment was repeated with other plasmids as well; they then furthered science by adding the genes from a toad to the bacteria. Astonishingly, the genes from the other organisms stayed vigorous in the bacteria they were added to.

Bacteria reproduce more or less every 20 minutes, thus, with these new findings, bacteria could now be used as a source of rapidly multiplying genes. The idea that one could add something, like the human insulin gene, into bacteria, let them reproduce every 20 minutes and have a large amount of insulin ready in no time, became a possibility.

The Stanford group of scientists, which included Berg, Boyer and Cohen, released a sequence of scientific writings that greatly changed the process of nature through the combination of genes from two different creatures; they successfully took the genes from one bacterium and implanted it into a different bacterium. Their work opened a whole new frontier in biology, which had been beyond even the realm of imagination a few years previously. The idea that a gene could be taken from a human cell, introduced into bacteria and made to function as it does in the human body was mind-boggling.

IN CONCLUSION

In this chapter, we noted several phenomenal advances. Enzymes were discovered that help remove molecules from one living being to be delivered into another, totally different, type of organism, where they functioned as they did in the original host. The human body consists of more than 300 trillion cells, making it truly multifaceted. A couple of months after researchers had inserted a gene from a frog into a bacterial plasmid, it became possible to recreate these results in a complex organism, as seen with the insulin example. Biology was transformed by these findings in very little time, opening up the possibility of crossbred bacteria designed to manufacture human proteins; or plant life with better nutritional value that can battle diseases; as well as creating many greater possibilities for humans.

SOURCES & REFERENCES – CHAPTER 20

Lear, John. *Recombinant DNA: The untold story.* Crown publishing, New York, 1978

CHAPTER 21:
BACTERIA & YEAST: THE NEW INDUSTRIAL WORKER

"Man's endless pursuit for finding nature's hidden treasures and secrets had rewarded him with countless discoveries in every field"

Unknown.

Colonel Eli Lilly was a versatile and active American industrialist in the 19th century; he was also the founder of Eli Lilly, the pharmaceutical giant. The Colonel adopted new approaches in his field and focused on research and developing new products, rather than simply manufacturing what they were already making. The company progressed and became a powerhouse in pharmaceuticals in the 20th century. It was the first company to manufacture polio vaccine and insulin in large quantities.

In 1978, Eli Lilly's plant in Indiana was using the pancreas, the organ that makes insulin, of 23,500 hogs and cattle to manufacture a pound of the hormone. The company, to meet the growing need of insulin in the country, utilized pancreases from 56 million animals per year. The predictions were dire that it might be impossible to meet the growing demand for insulin. Eli Lilly, the major producer of insulin in the US, was negotiating with leading research institutions in the country, including Harvard and the University of California San Francisco, to develop bioengineered insulin.

Genentech was a small start-up company in San Francisco, pioneered by biochemist Herbert Boyer and venture capitalist Robert Swanson. Boyer's knowledge of recombinant DNA technology to fuse human genes with bacterial genes to produce human proteins, together with Swanson's background in raising funds and management, showed the world how faith, determination and can-do spirit will overcome many obstacles. In 1978, Genentech had only 12 employees, when it decided to join the race to develop an alternative source of insulin that did not require sacrificing an animal to obtain its pancreas. It would involve introducing the human insulin gene into bacteria, which can reproduce rapidly and produce human insulin, which can then be extracted from the bacteria.

Swanson was told by one of his scientists that it could not be done on an industrial scale. Swanson did not want to hear, "it cannot be done". He asked, "What do you need to get it done"? Swanson raised more money; he recruited a team and remodeled an airfreight warehouse in San Francisco into a lab. The team went into action right away, working literally round the clock, with a day and a night shift. Swanson's energy and Boyer's technical expertise became the little David, competing with the Goliaths of academic institutions of the like of Harvard.

Amazingly, within a year, they successfully produced enough recombinant insulin to perform initial testing. Eli Lilly joined with Genentech and applied for six patents for different ways of creating human insulin through the use of recombinant DNA technology. Preliminary studies using recombinant insulin were performed at Guy's Hospital, in London, on 27 healthy volunteers; these studies proved the product was harmless. Further testing over the next four years confirmed its safety and efficacy, before the FDA (Federal Drug Administration of the US) approved it. The New York Times of 30th of October, 1982, on its front page, carried the good news of this phenomenal scientific achievement; a new human insulin product, known as Humulin, which eventually replaced completely the insulin extracted from animals.

CREATING HUMULIN

Humulin was the first-ever recombinant-DNA-technology-produced drug that was approved for use by humans. The source of Human Insulin is a commonly-found, innocuous bacterium in the human gut; E. coli. As such, the first recombinant DNA technology introduced a new age of development in medicine.

Diabetes, a common disease in the USA, with an estimated 9% of citizens suffering from it, is also one of the main causes of death. Individuals with Type I diabetes, meaning that their bodies do not make insulin, as well as some that suffer from Type II, (when the body is either resistant to insulin or it does not create enough), require insulin to control their disease. Banting, Best and McLeod were the three scientists, hailing from Canada, who first isolated insulin as a small protein in the pancreases of animals.

Lilly's primary facility for the manufacture of insulin in the USA was situated in Indianapolis until the mid-1980's. Midwestern pork farmers would send animal pancreases in great masses via trains, from the abattoirs to the structure adjacent to the tracks, where they would then mine the insulin from the animal's pancreas. The insulin mined was similar to that of the human body and, thus, well tolerated by most users.

Yet, there were some who developed hypersensitivity to the alien protein used within their bodies. Another main problem scientists encountered while mining the insulin from animals, was that it was only possible to gain an insignificant quantity of insulin from each animal. In one year, one individual used the insulin gained from almost seventy pigs. The productivity and sustainability of the newfound science was in question, especially due to the growing number of diabetes patients around the globe, estimated to reach 366 million by 2030. During the late 1970's and early 1980's dreadful forecasts were made for the lack of insulin in coming years, as, at that time, only livestock supplied insulin. Luckily, when *Recombinant Human Insulin* became available, things became easier; the insulin was now human, removing hypersensitivity, and could be mass-produced, eliminating the risk of shortage.

The patents and successful manufacture of recombinant human insulin tremendously boosted the profits of Eli Lilly and Genentech. Lilly built two plants to manufacture Humulin, in Speke UK, and in Indianapolis US, in 1980, costing around forty million dollars. The worldwide sale of *Recombinant Human Insulin* reached twenty-seven (27) billion US dollars in 2015, with estimates for future sales to grow to around forty-three (43) billion US dollars. Lilly is listed as one of the five main diabetic medicine providers across the globe.

VITAMINS

In 1920, a new group of molecules were found, with the first one named as Vitamin A. Vitamins are nutrients necessary, in small quantities, for the human body to ensure development, reproduction and overall health. But the body does not create vitamins. Vitamin A is mostly located in foods derived from animals, since it is fat-soluble. Worldwide estimates calculate that more than one billion people suffer from a lack of micronutrients, especially Vitamin A. This deficiency is one of the main causes of childhood blindness, and it is a huge problem in Africa and Southeast Asia. Since a large part of the world's populace only has access to rice as a source of food, especially in countries with low-income rates, they suffer since rice does not contain any Vitamin A.

Nowadays, in advanced countries, many types of foods, such as cereals, are enhanced with vitamins and other nutrients the human body requires. A deficiency of micronutrients is rare in the US, because of directives enforced by supervisory organizations monitoring nutrition. Countries that are poor have to rely on products that are easily available locally; the World Health Organization of the United Nations, along with multiple governments, has had limited success in solving this problem of malnutrition.

RICE

Recombinant DNA technology made it possible to advance the value of harvested grains like rice by adding new genes; these new genes would then create the necessary micronutrients, just as bacteria created human insulin. Rice made the best object for bio-fortification, since it was a prevalent crop in countries where there is a lack of Vitamin A. Almost seven years were spent on research by different teams and it finally became possible to add three genes to a rice plant; two genes came from daffodils and the other from a bacterium, which added a precursor of Vitamin A, *beta-carotene to the rice*. Further research allowed the replacement of one of the daffodil genes with one from the corn plant that enlarged the harvest by twenty-three times. This rice strain, known as GR2 or golden rice, can stop Vitamin A-deficiency-induced vision loss in children by using only a third of a cup a day. There are now multiple crops that have undergone this bio fortification, such as corn with Vitamin E and rice with increased iron.

IN CONCLUSION

Today, there are a wide variety of medications available that stem from advances made in DNA research, such as hormones, vitamins, clotting factors, and vaccines. This technology has transformed the making of medications and has opened limitless possibilities for better healthcare across the globe. There are even bacteria that have been engineered to clean oil spills. Recombinant DNA technology has begun an upheaval in engineering manufacture, adding bacteria, yeast and biological molecules and medicinal forms that are not derived from animals or chemicals. A new and efficacious source of labor, compliant, inexpensive and available in unlimited numbers has been discovered for the benefit of humanity.

SOURCES & REFERENCES – CHAPTER 21

Hughes, Sally Smith. *Genentech: The beginnings of biotech. University of Chicago Press*, 2011.

The Golden Rice Project. www.goldenrice.org/

CHAPTER 22:
GAIN BECOMES PAIN AS THE CLOCK TURNS

"The world moves, and ideas that were once good are not always good."

- Dwight Eisenhower

Mosquitoes and the malaria parasite have pestered humans since ancient times. In India, there are early Vedic writings explaining a fever, called 'king of diseases', which is similar to malaria. During the 1900's, between 150 and 300 million people died of malaria. However, as serious as the malaria parasite was, it was not one of the leading causes of death.

DOCTOR TONY ALLISON

At age ten, the son of an English farmer based in Kenya, became sick. Tony Allison had contracted malaria – an illness caused by a single-cell parasite. Due to the sickness, young Allison was changed for life, and thereafter had the desire to become a biologist. Yet, as often happens, life takes us on a different course; Allison pursued medicine instead.

After he finished his undergraduate degree in South Africa, Allison went on to study medicine at Oxford. He became interested in genetics and thought it would open a new door to understanding, not only diseases but also human history.

He had an opportunity to join a summer survey team in 1949, which was travelling to Kenya and other regions of Africa to study plants and insects. Allison joined the team to study blood groups and see if they would reveal the genetic relationship between various tribes. He sampled blood from multiple communities. In his search, he also looked for sickle hemoglobin, which exists in Sickle Cell patients to see if there was a relationship with any particular tribal group. Allison knew that Sickle Cell disease was a widespread hereditary malady in Sub-Saharan Africa.

Unfortunately, there were no real variances to be found in the blood group analysis; on the other hand, the sickle hemoglobin information revealed something extraordinary. Differences existed in the sickle hemoglobin analysis between the communities living on the coast or close to Lake Victoria and the communities in the plateaus and highlands, or dry areas. The samples from communities situated closer to water was over 20% positive with the sickle cell trait, while the samples from the dryer areas showed less than 1% positive. It was known, by then, that mosquitoes and malaria were prevalent in wet and coastal regions, while arid regions and plateaus had almost no problem with either. Was this the connection he was looking for? Could the sickle cell trait help us to find a way to defend people from the malaria parasite?

At the time, Allison was only a student in medicine and still had to go back after the survey to finish his degree. One year after he finished his degree, Dr. Allison gained financial backing and returned to study the relation between malaria and sickle cell in great detail. He collected blood from more than 5000 people across Sub-Sahara Africa. His study demonstrated that the sickle cell characteristic ranges from 0 to 40%. His studies confirmed the notes he made during his tour - sickle cell traits were widespread in regions where malaria was rampant, the wet lowlands, while it was almost completely absent in the regions where malaria was not found, the dry highlands

SICKLE CELL DISEASE

Chicago, 1910. Dr. Herrick, a cardiologist, receives a patient - a young African man from Grenada, who is suffering from chest and body pains; the young man is hospitalized where his file is given to the doctor on duty, Dr. Ernest Irons. Irons records that the patient is very pale and does an inspection of his blood under the microscope, where he discovers strange 'sickle'-formed red cells.

Herrick then described this as a potential new illness in a medical journal. In Africa, the presence of sickle cell disease is estimated to have existed for around 5,000 years, although it was not known as such. After Herrick's publication, more and more accounts became apparent in US hospitals, all with patients hailing from African lineage. Up to that time, the disease appeared to be inherited from the family, signifying a hereditary root. But things changed when a different case appeared: disease did not appear in the father of the patient, although the cells sickled in the laboratory. Not long after these appearances, British and French doctors reported diseases like the sickle cell in Africa and India; and later it was found in Greece and Italy as well.

The red blood cells are the oxygen carriers in the body. They are packed with a protein called hemoglobin that binds oxygen as they travel through the lungs, which are rich in oxygen when we breathe. As they travel through the body after leaving the lungs, they release the oxygen in the tissues, so that cells can turn food into energy for driving the metabolic processes of life.

At the University of Indiana, Hahn and Gillespie researched the red blood cells from a patient with sickle cell disease and found that the cells had become misshapen – the red blood cells changed from the original disc form into sickle form when they were put into a mixture of carbon dioxide; then, when they were put back into an oxygen mixture, they once again became disc-shaped.

Within twenty years, the hereditary nature of the sickle cell disease was well recognized. When an individual gains this gene from both parents, the full-blown illness develops, but if the gene is only gained from one parent, the illness does not develop but the individual becomes a carrier instead. This is called a recessive gene, which only manifests if both genes are abnormal; otherwise, the one normal gene from the healthy parent is able to mostly compensate and prevent the development of illness. However, some abnormal genes can be dominant, which means that inheriting a single gene from one parent, even if the other parent's gene is normal, leads to disease. Huntington's disease, which is characterized by uncontrolled body movements and dementia, is an example of a dominant gene.

This was the clarification for the unusual recorded case of sickle cell disease mentioned before; the parents were carriers, so they did not manifest the disorder, but the unfortunate child got two abnormal genes, one from each parent, and thus had the illness.

Linus Pauling, a double Nobel Laureate and extraordinary chemist, examined the hemoglobin molecules found in sickle cells in 1951, and compared them to normal hemoglobin molecules. Pauling had predicted, as early as 1945, that the two molecules would be different, and he subsequently proved that they were chemically different. He suggested naming this a 'molecular disease', so as to explain the cause as a transformed or anomalous protein structure. Vernon Ingram, a Cambridge University biochemist, furthered this research and found that only one amino acid was different between the sickle and normal hemoglobin. This apparently small difference, an amino acid called valine instead of glutamic acid, induced such a profound change in the molecule, causing a grave disease.

It is believed that this mutation happened around 70,000 -150,000 years ago. Children who gain the mutated gene from both parents, called the homozygous state, hardly ever live past the age of five in under-developed countries that do not have the facilities to care for them; most of the time they perish early from pneumonia. In advanced countries, however, the mutation can be identified at birth and steps can be taken to avoid the infectious complications of disease through the use of vaccines and prophylactic antibiotics; the mortality rate for children has, in this way, been lowered to less than 2%. Those born with only one of these genes, being heterozygous, have a normal life expectancy. An approximation made in 2010 showed that there are over 100 million sickle cell gene carriers; a precise approximation of the number of people with the disease is unknown, as there is not enough information in countries where the occurrence is widespread, particularly because of the early mortality rate being so high. In 2010, there were around 100 thousand cases in the USA.

SICKLE CELL AND NATURAL SELECTION

Darwin's principle of natural selection, if correct, meant that this dangerous mutation should have died out over the years. Those with only one of these mutated genes draw the benefits, though; a heterozygous condition creates a reasonable amount of defense against the malarial parasite, making it harder for the parasite to enter the red cells, making the disease less deadly for them. Worldwide, homozygous sickle cell disease has perhaps taken the lives of 100 thousand children, but malaria has taken 1.5 million; thus, the sickle gene has been conserved by natural selection, as the heterozygous state protected against malaria.

Tony Allison had discovered precisely this while conducting research in Kenya. Regions that had widespread mosquitoes and malaria also had widespread heterozygous sickle cell characteristics, as a mutation giving some defense against the deadly disease. Regions that were not rampant with malaria did not have this mutation because it was not useful there. In industrial countries, infant deaths have been eliminated; thus, people with sickle cells suffer later complications from the illness, which include episodes of severe pain. What was an advantage in the past has become a pain, because early deaths from infections in the homozygous kids have been practically eliminated by advances in medical technology.

V LEIDEN MUTATION AND NATURAL SELECTION

After trauma or an operation, a complex process follows for blood to clot; this process comprises multiple enzymes and proteins working together to make the clot. One single error during any part of the process can easily cause either bleeding, such as hemophilia, or excessive clotting, known as thrombophilia. Factor V Leiden is a fairly well-known mutation in the Caucasian population. It likely happened millennia ago, in one person in Northern Europe. In the United States, of the population that descended from North Europeans, 3% have this mutation today.

An everyday reason for the death of young females in the past was infection or bleeding after delivering a child. The Factor V Leiden mutation lessened the danger of bleeding and, possibly, the danger of infections as well; this mutation thus carried great benefit. A disadvantage, though, for carriers, lies in the greater tendency to have blood clots in their legs, which can be deadly if the clots migrate to the lungs. In the past, the fertility benefits outweighed the dangers of clots, as life expectancy was short, while blood clots mostly developed in older age. Natural selection augmented the number of people with the mutation so much that it is now one of the most common mutations in the North European Caucasian population.

However, advances in medication to control the bleeding and infections from labor negated the advantage, while longevity has brought pain in the form of a growing danger of blood clots for those who harbor this mutation.

IN CONCLUSION

Today, medication that can decrease the danger of malaria and childbirth bleeding has made some previously beneficial mutations harmful. In the past, natural selection would have eliminated them, as infants born with homozygous sickle cell mostly died in infancy. In modern times, such infants in developed countries survive to adulthood and bear children, thus perpetuating the survival of this gene. The carrier state of sickle cell was an advantage in protecting against malaria in the past, but with effective treatments for malaria and mosquito control measures, the gene is of little value but will persist in the population unless we can make it disappear, with genetic counseling to prevent two people with the disease or carrier state from marrying each other, or at least not having children together.

SOURCES & REFERENCES – CHAPTER 22

The discovery of resistance to malaria of sickle cell heterozygote. [Online] Available at:
https://iubmb.onlinelibrary.wiley.com/doi/full/10.1002/bmb.2002.494030050108

Lindqvist, P. G. "On the evolutionary advantage of coagulation factor V Leiden (FVL)." *Current medicinal chemistry* 22, no. 32 (2015): 3676-3681. Available online at: https://www.ncbi.nlm.nih.gov/pubmed/26423085

CHAPTER 23:
UNJUST INCARCERATION

"DNA technology could be the greatest single advance in the search for truth, conviction of the guilty and acquittal of the innocent since the advent of cross-examination."
- *Judge Joseph Harris*

July 9th 1977, near the Homewood Shopping Mall in Chicago, Illinois, a police officer sees sixteen-year-old, Cathleen Cromwell, next to the road wearing dirt-stained clothes. Cromwell, through her tears, tells the officer that she was at the Long John Silver Seafood Restaurant close by and while she was walking home from work, a car pulled up next to her. Two of the three young men jumped out and forced her into the car; she says one tore at her clothes, raped her and used a broken bottle to scratch letters into her stomach, after which they dumped her next to the road.

She was taken to the nearest hospital, where she underwent a rape test, and doctors found that she had, indeed, had recent sexual activity. The investigation began and police searched for the assailant. But did he really exist?

THE STORY

Cathleen grew up in a troubled family. When she was fourteen, she was adopted and went to live with foster parents. Finally, she had found a place where she could feel safe and where she was cared for. She fell in love as a teenager. She and her boyfriend took their relationship to the next level; but the day after they had sex, she panicked, fearing she might be pregnant and that her foster parents would kick her out. To try and resolve the issue, she came up with a farce and made up the story of being raped. Before the police officer spotted her, Cathleen had gone into the woods, inflicted bruises on herself, tore her own clothes, and used broken glass to carve into her stomach. The idea was to tell her foster parents that she had been raped, but she did not want to tell the police. However, the police officer found her right after she had come out of the woods. She was crying and upset, sincerely, but not for the reasons she gave the police.

Now she had a new fear: her parents would be angry with her for this deception, so she went ahead with the rape allegations. The police showed her mug shots of people similar to those she described, but she did not identify anyone; so, a sketch artist was brought in so she could make an in-depth description, while she hoped that no-one would match and the case would go cold. Unfortunately, a few days later, one of the police officers took a look at the sketch and thought that it could be Gary Dotson. Twenty-year-old Dotson was a high-school dropout and had been convicted of having a stolen television in the past. The police took his photo, along with others, to Cathleen's foster home and showed her a photo line-up. She, of course, denied seeing the assailant. The officers compelled her to take a better look and she saw a photo that looked like the person she described; fearing that the officers would find out she was lying, she identified Dotson. She once commented that, at that moment, she thought: "I hope this guy has a good alibi for where he was." Sadly, he did not, and, what was even worse, the evidence made him look guilty.

THE TRAIL

A forensic expert for the state police gave testimony that, in the semen stain on Cathleen's underwear she had worn that night, he had found type B blood antigens; Dotson's blood type was type B. He also stated that just 10% of American white males had type B blood, but he did not mention that Cathleen was also type B, or that there were type A antigens in the semen stain as well.

In his own defense, Dotson attested that he had attended a party that night; three friends stated, under oath, that he had been with them at the party. Other discrepancies also existed; Cathleen had said her assailant was cleanly shaved while Dotson had a mustache. But Cathleen claimed, "I can never forget that face"; she said. Dotson was found guilty and condemned to twenty-five to fifty years in prison. As officers took him away, he cried out, "I didn't do it!"

WHAT HAPPENED NEXT

In 1981, Dotson's family had finally put together enough money to make an appeal, but the appeal was rejected; the innocent Dotson's destiny appeared to be inevitable. During this time, Cathleen had finished high school, married her high-school boyfriend, and moved to New Hampshire. Here, she became devoted to Christianity, and the guilt became too much for her; she confessed her lie to her minister's wife and to her husband. He insisted that she take responsibility for her actions. She hired a lawyer and phoned the police. Before the new hearing, Dotson made bail and was released; then, on the 11th April 1985, Cathleen took the stand and admitted to her lies. She explained that her motivation came from fearing the loss of her foster parents if she were to become pregnant after she and her boyfriend had had sex. Judge Richard Samuels, who had been the judge during the original trial, rejected her new testimony, stating that her testimony was unreliable.

> *"The jury and I found her [1979] testimony to be credible. I observed her demeanor in court. It was the demeanor consistent with someone who had been raped [...] She had an inability to recall certain items and she was impeached on certain items."* - Judge Richard Samuels

Michael S. Serrill, then a *Time* Magazine reporter, covered the case and stated that, *"the courts have always regarded recanted testimony with suspicion; in part, because there are too many bad reasons for witnesses to change their minds: intimidation, bribery, misplaced sympathy for an imprisoned or condemned offender."* Thus, the verdict was not overturned and Dotson was once again placed in prison, despite Cathleen's objections. Dotson's lawyer appeared on CBS's morning show, making the case national news and resulting in a large amount of public sympathy. Over seventy thousand Illinois locals appealed to the then governor, Governor James Thompson, to take over the case and release Dotson.

Governor Thompson took the case and held a special hearing, along with the Illinois Prisoner Review Board, during which Mark Stolorow, the Illinois State Crime Lab's forensic expert, contradicted the previous forensic expert, Timothy Dixon. Stolorow said that Dixon misled the jury when he said that it was only possible for 10% of American white males to have deposited the semen on Cathleen's underwear. Stolorow argued that the semen could actually have come from 66% of the population, which made the chance that it was Dotson that much less. Dixon had also stated that there were three hairs found on Cathleen's body that were like Dotson's; Stolorow, however, stated that there was only one such hair, not three.

Dotson's lawyer said that Dixon had given a false testimony, allowing the jury to believe that he had graduated from the University of California, Berkley, while he actually had not. Dixon repudiated these allegations, saying that he attended an ascribed alumna session there; repudiating that he had said he had graduated from U.C. Berkley. Even though the forensic evidence was very convincing, it was not final; then, after hearing Cathleen's confession, just as before, Governor Thompson did not pardon Dotson. All he did was lessen the ruling to the six years he had been in prisoned, and Dotson was released on parole.

Dotson may no longer have been in prison, but he had not been exonerated, and he spent every day endlessly suspected of being a rapist. He struggled to find work and became an alcoholic; in 1988 he was in a bar fight and was sent back to prison for parole violation. Dotson would only be released if he could prove he did not belong on parole; for this to happen, he needed to be proven innocent. In 1988, Thomas Breen, Dotson's new layer, asked for a DNA test - something that had not been possible during the initial trial. Breen asked that Cathleen's underwear from that night be the object of the DNA test. His request was accepted.

THE DNA TEST

The Forensic Science Associates in Richmond and California did the DNA test and found that Dotson could not have been the donor. On the 15th of August 1988, Governor Thompson and the prosecutors heard that the DNA test had definitely disqualified Dotson; the test identified David Bierne, who had been Cathleen's boyfriend at that time, as the source of the semen.

Cook County's main judge ruled that Dotson was to be given retrial; but the State Attorney's Office did not want to impeach on the basis of the outcomes of the test or the victim's 'lack of credibility.' Dotson's sentence was overturned; in August 1989, Judge Thomas R. Fitzgerald held a quiet court and said to Dotson, *"You, sir, are discharged."*

Dotson was exonerated after having spent eight years imprisoned for a crime he did not commit, and four years after Cathleen had withdrawn her accusation. In 2002, Governor George Ryan formally absolved Dotson. At last, he was free, but it was not thanks to Cathleen or any lawyer; it was thanks to the science of DNA profiling.

DNA PROFILING: ALEC JEFFRIES' INNOVATION

Since the 1200's, Oxford University has been at the forefront among the learning centers of the world, famous for inspiring unconventional thinking and pioneering research. Alec Jeffries found this to be the perfect place for him. He was born on the 9th of January 1950, and, from an early age, was interested and curious about science. At 12 years old, he designed his very own dissection kit, which had a scalpel made from a crushed pin. When he was eighteen, on his birthday to be exact, he received a chemistry set and, from that point on, spent endless hours doing trials and making solutions; some of which sometimes caused explosions. In a few years, his home laboratory grew, after he was given a lovely Victorian brass microscope, which allowed him to do in-depth research on a variety of biological specimens.

In the 1960's, Jeffries was a teenager and socially identified with hippies; but, fortunately, he still loved science and went to Oxford University. He finished his undergraduate degree in biochemistry with first-class honors, after which he completed his PhD in genetics. Jeffries went on to the University of Amsterdam to study mammalian genes, and then returned to England, joining the University of Leicester, where his studies turned to tracing hereditary gene markers to recognize congenital illnesses.

Genes determine the traits of all living creatures. Genes are composed of macromolecules known as nucleic acids. The specific nucleic acid which carries the genetic material is deoxyribonucleic acid, or DNA. It consists of a series of nucleotides, also called bases, wound in a double helix manner. The base on one coil of the helix, pairs only with a specific base on the other coil. A base called Adenine will only pair with a base known as Thymine, while Guanine will only pair with Cytosine. The two coils are held together by chemical bonds, known as hydrogen bonds. The total number of bases in each cell is called the genome.

The order of these bases regulates everyone's genetic traits; every cell in the body has the same DNA. DNA can be found in smears of blood, drops of semen, skin cells, spit, and body tissue, even emissions such as urine or sweat. There are around three billion base pairs in the human genome, which made studying it very time intensive and problematic. Jeffries studied the mini-satellites, a part of the DNA sequences, which occur repeatedly throughout the genome. When a DNA coil is cut into fragments, using the molecular scissors, these mini-satellites can be identified. These sequences vary in length from individual to individual. This is also called restriction fragment length polymorphism (RFLP).

In a mini-satellite, often only 10 to 60 base pairs exist, but sometimes 100 or more may be present. These mini-satellites are known to exist at more than 1,000 locations in the human genome. No two people, except identical twins, share the same set of regions. This fact was unknown till Jeffries discovered it. All the nucleotide bases do not code as genes for a trait, but may have other functions.

With the hopes of creating a way to separate various mini-satellites at the same time, by making use of chemical probes to identify specific chemical patterns which categorized DNA areas, Jeffries X-rayed DNA. On September

10th 1984, in the morning, Jeffries was studying one of the x-rays and came to a surprising conclusion. He saw multiple mini-satellites, as he had expected, but also found that no two people had the same mini-satellites. It is much easier and less time consuming to study the restriction length polymorphism than map out the entire three billion base pairs in a person's sample and compare them to another persons. This was similar to having fingerprints which are different in every person.

> "It was purely by accident that we generated, totally out of the blue, what proved to be the very first DNA fingerprint [...] Purely by chance, we had a family group on there, one of my technicians and a mother and father. Not only could we tell those three people apart, but we could see how the technician's pattern was a composite of some of mum's and some of dad's characters [...] To my amazement, not only did this work on people, it worked on absolutely everything else as well. The penny dropped almost immediately. Within a minute I suddenly realized, wait a minute. We're looking here, potentially, at DNA-based biological identification." - Alec Jeffries

Jeffries instantly thought of the effects it would have for forensics.

> "Was it possible to use this in a case?
> Did the DNA from crime scene samples survive?
> Was it possible to tie the samples to a suspect?"

Another problem was also instantly solved, as a child's paternity could now easily be identified. One could simply tie the parent's DNA fingerprints to the child's, instead of trying to match blood groups. The genetic fingerprint was found by accident! Jeffries had made a groundbreaking discovery, changing technology to solve multiple issues, such as parenthood and migration disagreements, searching the possibility of birth mutations, recognizing remains; even animal breeding would benefit. He patented the technology.

From 1984 to 1987, he and his team had the only laboratory capable of finding DNA fingerprints worldwide. After 1987, he made the technology accessible for commercial use and since then it has been a foundation for forensics in law. His work was able to identify perpetrators from any form of body fluid, skin cells or hair found at a crime scene; and, as in the case of Gary Dotson, this technology was a way to prove whether a person had been incorrectly suspected or unjustly sentenced. This spurred on a group of Yeshiva University's law students to begin the Innocence Project in 1992, by making use of DNA testing to rectify miscarriages of justice of the past.

IN CONCLUSION

The National Research Council conducted a study for the United States Department of Justice in 2006 and found that the leading cause of over 70% of unjust sentences in the United States was eyewitness misidentification. Today, the court system values DNA fingerprint science over witness accounts that could be biased in many ways. The Innocence Project stands up for those who were unjustly sentenced and are constantly advocating changing the law to allow for more DNA testing for closed cases. More than 350 people, in the United

States alone, have been exonerated because of DNA testing since 1989, after they were sentenced for crimes they did not commit; 20 of these unjustly sentenced people were sentenced to life in prison. DNA testing revealed the true criminal in 160 cases; thus, DNA testing gave justice to, not only the unjustly sentenced, but the victim of the crime as well.

SOURCES & REFERENCES – CHAPTER 23

Profile of Alec J. Jeffreys. PNAS (Proceedings of the National Academy of Sciences of the United States of America). [Online] Available at: https://www.pnas.org/content/103/24/8918

Fridell, Ron. *Solving Crimes: Pioneers of Forensic Science*. Franklin Watts, 2000.

CHAPTER 24:

ULTIMATE GENEALOGY: TRACING THE ROOTS OF ADAM & EVE

"To forget one's ancestors is to be a brook without a source, a tree without a root."

- *Chinese Proverb*

Anthropology has always been a science that makes use of a large variety of methods to research and track the ancestries of populaces; anthropologists use languages, corporeal factors, ancient archives, geophysical aspects, utensils, trinkets, ceramics, holy signs, and so on to find ancestries. When DNA technology appeared on the scene, the arena of anthropology was completely transformed forever.

THOR HEYERDAHL AND THE CURRENTS

The only son of a Norwegian family, Thor Heyerdahl, was born on the 6th of October 1914, in Larvik. During his early childhood, he often accompanied his mother to the museum since she was the director. At the museum, young Thor preferred to learn about Darwin than listen to fables. Throughout his childhood, adventure called to him; he even went on a hike once with a dog, so that he could sleep in the snow. His intention was to show that he could take the weather all on his own.

In later years, he received degrees in Zoology, as well as Geography, and journeyed to Fatu Hiva – an isolated island in the Pacific Ocean, along with his new wife. There, he conducted research on indigenous wildlife during 1937. They and the locals became quite close, so much so that they embraced the style of living and the tribe's high chief adopted them. Heyerdahl was there to study the transoceanic beginnings of the island's animals; but his interests shifted to the people, changing from Zoology to Anthropology.

More or less 3,000 years ago, the population of these Pacific islands now known as Polynesians, began settling there. Heyerdahl wanted to know where exactly these people came from, as the textbooks of that time simply said they had come from Asia. He had found that a few of the plants on the island, such as the sweet potato, originated from South America. Another observation he made was that the stone constructions found on the island were also very similar to those in South America. One day, while on a fishing trip, he noticed that the ocean currents went east to west, which made him doubt the reputed heritage of the island's population.

The Magellan expedition, the first exploration of Pacific by Europeans, which could not sail back east after crossing the Pacific as the trade winds are east to west, hence it continued sailing west and reached back to Spain circumnavigating the globe. Heyerdahl could not comprehend how, many eras before the Spanish, Asians had made the thousand miles long journey against the Pacific currents to the islands on paddleboats to inhabit Polynesia. This persuaded him to think that the inhabitants of Polynesia originated in the Americas, not in Asia as the history books had told.

He went back to Norway to investigate his theory further; he theorized that Peruvians traveled to the island in the Pacific using the natural Humboldt current of the South Pacific. Unfortunately, the Second World War disturbed Heyerdahl's studies, and he became a military parachutist.

After he returned from the war, he sought to verify his theory by building a raft, like the ones the earliest Indians of Peru had, and sailing to the islands. He, along with five of his friends, set out and constructed an indigenous-style raft, using wood like the wood that would have been available to the earliest Indians, as well as the tools they would have had; a 40-foot raft, named after the Inca god Kon-Tiki, with a bamboo cabin, which had a banana-leaf roof, sails and paddles.

For the journey, they were only going to use a radio and did not take navigation gear. Instead, they navigated by making use of the sun, stars, wind and currents to navigate the ocean.

Heyerdahl and his team set out from Callao, Peru on the 28th of April in 1947. During their journey, they lived off fish they caught from the boat's deck. This excursion was dangerous, filled with storms and great white sharks, and not many people thought they would make it. But 101 days after setting sail, they reached an island near Tahiti, an almost 5,000-mile journey. Heyerdahl verified that this feat was possible, and, at the same time, that the settlers of Polynesia could have been from Peru.

DNA AND ANTHROPOLOGY

DNA is present in the nucleus of the cell, but some DNA is found in the organelle Mitochondria - the part of the cells that make energy. The DNA found in the nucleus is gained from both the father and mother, but the DNA in the mitochondria only comes from the mother of the child. The amount of DNA in the mitochondria is very small, so it is easy to analyze; meanwhile, the DNA in the nucleus is a mix of both parents, who, in turn, received DNA from both their parents, so the hereditary line becomes intricate and hard to focus.

As mitochondrial DNA only comes from one's mother, who received it from only her mother, the possibility of recording and finding the origin of one's maternal ancestry becomes real and can hypothetically be done; unfortunately, in truth it is not so simple, because many unplanned mutations happen over the span of generations, and these mutations are then transferred to the next and so on. The rate of mutations can, however, be estimated fairly accurately.

Linking the mutations can track the associations of the people from their maternal ancestry, while the number of mutations can approximate the time over which they occurred. When mitochondrial DNA analysis was performed on Polynesians, it unequivocally revealed South East Asian lineage, instead of relocation from South America as Heyerdahl was proposing.

THE ICEMAN OF THE OTZAL MOUNTAINS

September 19th 1991, a German couple was hiking on the Austrian-Italian border in the Alps, when they came across something very bizarre; the upper part of a body was sticking out of the snow. When they returned from the hike, they reported what they saw and police were on the scene the next day. They first thought that the body was that of someone who had an accident on the mountain, dying in the snow. When they tried to remove it, however, they found that the lower part was completely enclosed in the ice, with an axe nearby.

On September 21st, two experts arrived at the scene, and on examining the body dressed in leather clothes and with an axe so close by, their first thought was that it was quite old. After the body and associated items were recovered, they were taken to the Department of forensic medicine at the University of Innsbruck, Austria. The items found next to the body consisted of a dagger, some pieces of leather, hay clumps, a long stick and the axe.

On the 24th of September, Konrad Spindler, from the University of Innsbruck, who had knowledge in the fields of history and prehistory, was asked to assess what had been found. Instantly, he asserted that the mummy was a minimum of 4,000 years old. This finding rattled the entire globe, since it was the oldest mummy ever found in Europe, and must have been alive approximately 3,300 BC. The mummy was dubbed the Iceman, or Otzi because it was found in the Otzal Mountains, and is now on display in Bolzano, a city in Italy.

BRYAN SYKES AND OTZI

Bryan Sykes was a genetics professor at Oxford University, and a pioneer in the analysis of DNA from ancient bodies. DNA does not degenerate over time and natural factors do not destroy it easily; thus, Sykes could gather DNA from Otzi, and analyze the mitochondrial DNA. The findings startled him. Otzi's mitochondrial DNA was very akin to the mitochondrial DNA of a management consultant in Southern England, which was astonishing! The 5,000-year-old Iceman's relation was alive in Dorset, England.

This data started Sykes' studies in the hereditary origins of the European populace, using maternal DNA examples from the continent. He found that 95% of the European populace could be put into seven groups, or haplotypes of mitochondrial DNA. His model showed that he could trace the ancestry of 95% of Europeans to just 7 women, who were not related to each other. Thus, almost all Europeans would be related to one of these seven 'Eves,' the first mothers – or clan mothers as Sykes would say - in each family, who were alive somewhere between 11 and 45 thousand years ago; researchers later found that there were possibly even more haplotypes. It is possible that these haplotypes could be related and connected to each other, except that there are no surviving relatives to connect the dots between these groups.

Other continents have different "clan mothers"; mitochondrial DNA information has been taken around the world and this has found that there was a shared maternal predecessor for everyone on earth. She has been named MRCA – most recent common ancestor: she is where every line of maternal ancestry from the populace of the world today joined; our genetic or biological Eve. The estimates for our greatest grandmother vary from 100 to 200 thousand years ago. Strictly speaking, this does not mean she was the only woman; merely that she was the only woman whose offspring lived on to today.

BUT WHAT ABOUT THE MEN?

Males also have a very different chromosome, namely the Y chromosome, passed on from their fathers. Like the mitochondrial DNA, the Y chromosome can be used to trace the male ancestry. The research team at Stanford University gathered and analyzed the DNA from 69 men from all over the world; this study showed that all the men tested had a mutual male forebear, between 125 and 156 thousand years ago. Other such research proposed that the mutual male forebear was alive almost 200 thousand years ago.

HAS SCIENCE TRACED ADAM AND EVE?

Do not, however, think that the mutual male and female forebears were the biblical Adam and Eve; they were merely the only two people whose offspring have survived until the present day. The other clan mothers and fathers or their offspring unfortunately perished without children to continue their genetic lines.

It is also likely that the genetically traced clan mother and father had no contact with each other, as we are tracing back from present day survivors who could be from different mothers and fathers. The dates estimated for the time when these clan parents lived vary from different studies, based on the methodology used. It is very likely that future studies will provide clarity to our understanding of the beginning of the human species.

CONCLUSION

The search for biological Adam and Eve will continue as more information becomes available. The genetic data, the archaeological findings, the fossil records and all other discoveries will hopefully, one day, be possible to match to visualize the biological Adam and Eve, their life and the world in which they lived.

SOURCES AND REFERENCES – CHAPTER 24

Sykes, B. *The Seven Daughters of Eve: The Science That Reveals Our Genetic Ancestry.* W. W. Norton & Company 2005.

"Genetic 'Adam and Eve': All Humans are Descendants of One Man and Woman Who Lived Over 100,000 Years Ago". *Ancient Origins: Reconstructing the Story of Humanity's Past.* [On https://www.ancient-origins.net/news-evolution-human-origins/genetic-adam-and-eve-all-humans-are-descendants-one-man-and-woman-who-021536

CHAPTER 25:
SEARCHING FOR LUCA (LAST UNIVERSAL COMMON ANCESTOR) IN THE RNA WORLD

"Maybe you are searching among the branches, for what only appears in the roots."

- Rumi

Nobel Laureate, Harold Urey, best known for his work on radioisotopes, was conducting a seminar at the University of Chicago. In attendance was young Stanley Miller, his doctoral student. Urey's seminar was about the beginnings of the solar system and the theory that there was potential for the atmosphere of primeval earth to sustain the production of biological molecules. Miller was intrigued by this concept and asked Urey permission to study the idea further. Urey, however, had his doubts that it could ever be proved, since this type of experiment had never been done in a laboratory. But Miller was determined and urging, and eventually Urey gave in and approved. As such, they set out to create a way to study the beginning of life.

At the time, they thought that primeval earth consisted of water, methane, ammonia and hydrogen in the atmosphere, so they mixed these substances into a flask and used electric sparks to recreate lightning. Miller was thrilled when the analysis of contents of the flask revealed five amino acids - the fundamental units of proteins. Their discovery was published in the Journal of Science of May 15th 1953, which would cement Miller's fame. Later, he was presented on the cover of Time magazine.

Miller and Urey had made the "primordial soup" that J.B.S. Haldane, a British born scientist, who had made India his home, had suggested in the 1920's. This created a novel area of research for scientists to study the processes nature uses to create life.

In later years, scientists, by using better equipment and advanced methods, found more than just the five amino acids that Miller had found. The story was very believable: a new planet whose atmosphere contained many gases, one lightning strike and boom, a volcano erupts – very similar to what happened in Miller's flask; but it was a lot more complicated. Recent research shows that earth's atmosphere was not what Miller had thought in the study; it also did not create self-duplicating molecules or a basis of ongoing power to allow the devices of life.

LUCA

Approximately 3.6 billion years ago, more or less, LUCA, also called LUA, last universal ancestor began its life span – or so say the scientists. It was almost certainly no more than a single cell organism, though; a little bundle of the draught molecule DNA. LUCA was supplied with the needed chemicals to survive and duplicate; stromatolites are edifices that can be around 4 cm large, created by groups of bacteria, which are capable of trapping sunlight and can use this for power to duplicate. This is similar to photosynthesis, the process plants and bacteria use to manufacture starch from a mixture of water and carbon dioxide. The earliest known stromatolite was discovered in 2012, in the shallow seas of Greenland, but it was thoroughly studied and published only in 2016. This stromatolite was thought to be 3.7 billion years old. Could this, perhaps, be the LUCA? The process of natural selection and the Darwinian progression became the leading factors in its development hereafter.

We learned from the Miller-Urey studies that basic chemicals, such as amino acids, could have resulted from an unstable atmosphere on the primeval earth; and that the chemicals that form LUCA are the information needed for understanding the beginnings of life. Multiple scientists have struggled to further this point.

Before the 1800's, scientists thought that chemicals in living creatures were different and came to be because of a "vital force." This was known as the theory of "Vitalism". It came from the idea that every living being needed a soul that provided it with life, in contrast with a rock that contained different chemicals and had no soul to bring it to life. Thus, the concepts developed of organic - the chemicals found in living creatures, that were alive - and inorganics, the chemicals found on things that were not alive. Friedrich Wohler, a German chemist, using inorganic chemicals, made Urea in the lab in 1928; Urea is a chemical found in the urine, composed of carbon dioxide and ammonium sulfate. Later, it became possible to recreate a multitude of organic chemicals in labs; organic molecules have even been discovered in meteorites, which crash on earth from other stars and planets, as well as in stardust. The fact that organic molecules were found on meteorites and stardust ignited the idea that life did not begin on earth but came from space, possibly another planet. Today, it is well known that organic composites are not the products of life but the outcome of steady chemical reactions.

THE FIRST MOLECULES OF LIFE

In 1977, organisms were found thriving on the seabed close to hot springs. The organisms did not receive rays from the sun and were under constant impact from a mixture of hydrogen, carbon dioxide and sulfur that came streaming out of the nearby geyser at very high temperatures. A large number of scientists thought that this might be where the first molecules of life had originated. How these molecules are created is still unknown to us, even though they exist in every living cell.

Even in science, there is argument about what came first - the chicken or the egg - the protein or the nucleic acid. Another big problem is that every organism alive today has these two molecules that are dependent on one another to survive. The simplest way to think of this would be to envision them being made concurrently, but this seems to be very improbable; the other method would be that one of the two performed every role. RNA seems to be the most likely agent, since it can be found in every cell, and it is the main data tool used by the cells to interpret the gene plan and enable the creation of proteins.

Woese, Crick and Orgel suggested, in the 1960's, that life on our planet grew round RNA and that it was later trailed by protein production; it was widely famous as the RNA world hypothesis. RNA was created in the primordial soup, then raised itself to create proteins and intricate organisms, and is now global and adaptable.

Another theory is that metabolism or metabolics came first; it suggested that carbon-based molecules came into existence first, such as acetate – a dual carbon mixture consisting of water and carbon dioxide. Then, other reagents, such as natural resources or permeable rocks, allowed for the creation of intricate molecules in thermodynamically promising conditions; some of the more intricate molecules could then be a reagent used to create even more intricate molecules.

But this did not explain the ability to self-reproduce. Yet another argument is that RNA reproduction is the second stage, meaning there was a stage even before that; then the third stage was protein production, after which came the development of DNA.

New studies propose that clay played an important role in the beginning stages of evolution of the molecule of life. Clay can be described as a sponge; it takes in the chemicals found in water, if allowed by perfect conditions, and these chemicals are then able to react and form intricate molecules, particularly since clay can be a reagent as well. The clay can act as a membrane to the molecules; scientists think that clay originated at more or less the same time as life. This links the biblical notion of creation – man was made from clay, and the Egyptian notion – man was made from silt and Nile water - to science.

IN CONCLUSION

Miller and Urey's study began a race in the abiotic chemistry area – the creation of molecules usually located in living organisms; this field created amino acids, but also sugars and nucleic acid bases.

Lately, even genes have been reproduced, although some steps use enzymes found in nature. However, these man-made genes only work when they are presented to a living cell. The amount of success that science might reach when trying to reproduce the unique chemicals that began life is still only conjecture at the moment.

SOURCES AND REFERENCES – CHAPTER 25

Gargaud, Muriel, Ricardo Amils, and Henderson James Cleaves, eds. *Encyclopedia of astrobiology*. Vol. 1. Springer Science & Business Media, 2011.

Weiss, Madeline C., Martina Preiner, Joana C. Xavier, Verena Zimorski, and William F. Martin. "The last universal common ancestor between ancient Earth chemistry and the onset of genetics." *PLoS genetics* 14, no. 8 (2018): e1007518.

CHAPTER 26:
ORPHANS IN A RICH KINGDOM: VIRUSES

" Many key concepts concerning the nature of immunity have originated from the very practical need to control virus infections"

- *Peter C. Doherty*

The mummified body of Ramses V, an ancient Egyptian pharaoh, was closely studied. Interestingly enough, it was found that Ramses V had smallpox. Smallpox is an illness that plagued mankind for ages. There are even accounts from 1122 BC China that explain the illness - as well as scriptures from Ancient India. Smallpox affected the Romans, contributing in some part to the deterioration of a once-powerful empire. Europeans brought the disease to North America.

Four hundred thousand people perished annually because of smallpox during the 18th century. The good news was that if someone had suffered from the disease and survived, they then became immune to it. An interesting discovery made then was that, if a little bit of fluid from the abscess of an infected person was presented into a healthy person via a small cut in the skin, that person would gain a certain amount of resistance to the disease. This procedure, dubbed variolation, was done in Africa, India and China and it was introduced to England through the interaction they had with the Ottoman Empire. Charles Maitland obtained a Royal warrant on the 9th of August 1721, to begin tests of variolation on six convicts; these convicts were assured the "favor of the king" if they were to do the tests. All six convicts survived the tests; not only that, they had become immune to smallpox. After this successful test, the procedure became widespread.

SMALLPOX AND COWPOX

On the 17th of May 1749, Edward Jenner was born in England. Jenner is known as the Father of Immunology and credited with the development of the smallpox vaccine. When he was a child, a dairymaid had said to him that she would never have smallpox as she already had cowpox; it was widely known that dairymaids, in English villages where dairy farming was the main source of income, never contracted the smallpox disease.

When Jenner grew up and became a doctor, he wanted to study and understand the connection between smallpox and cowpox. He later tested the theory of cowpox spreading from one individual to another, and how the infection could then protect a person against smallpox. He vaccinated a boy of eight years with excretion from a sore of a dairymaid with cowpox; then, in two months, he presented fluid from a sore of someone with smallpox to the boy and the boy stayed healthy. Jenner proceeded to repeat this experiment with multiple persons. Once approved, the method became the norm as protection from smallpox.

A main accomplishment of contemporary medicines is the fact that smallpox has been eliminated from the world by general immunization. Though people knew that it was an illness that was contracted by one person from another, the root of the illness was still unidentified; microscopic studies of the excretion showed no noticeable elements.

One of the great scientists, Louis Pasteur, saved countless lives by developing the method of pasteurization, the disinfecting methods for operations and inoculations against rabies and anthrax. He also verified and advanced the germ model, which explains that bacteria are the reason for rotten food and infectious sicknesses. One of his lab aides, Charles Chamberland, created a filter from unglazed porcelain in 1884. It gave water, when filtered through it, that had no bacteria in it, for Pasteur's studies. This contraption filtered bacteria, which were, at the time, the tiniest known living organism; anything smaller than bacteria could, unfortunately, go through.

TOBACCO VIRUSES

A sickness that made tobacco leaves mottle was known to tobacco farmers; until today, this sickness is the cause of the death of 2% of North Carolina's tobacco crop yield. In 1892, a Russian scientist received tobacco leaf extract that was mottled; he put it through the Chamberland filter and found the fluid that had been filtered still made the plant sick. He deduced that it must be a poison causing this.

But Martinus Beijernick, a Dutch scientist, did a similar test in 1898 with extracts that did not have the sick leaves; he found that the sickness could be spread by the filtered fluid and named it, *contagium vivum fliudum* – contagious living fluid. He suggested that the infectious agent was in liquid form when it went through the Chamberland filter; this then became known as a "filterable agent". The word "virus" was first used for filterable agents that potentially caused sickness; but is now used only for filterable agents that need a host that is alive and facilitates its replication and spread. A virus is around one hundred times smaller than a bacterium, so tiny that a normal microscope cannot see it. Viruses first became visible to us in 1930, when electron microscopy was developed.

THE FIRST VIRUS

Yellow fever was the first virus found that caused people to become sick; a pathological sickness spread by mosquitoes. The illness can be insignificant and self-restrictive, but it can also be stark and create deadly problems. In 1793, Philadelphia was struck by a yellow fever epidemic, with 10% of the citizens dying as a result.

Yellow fever was found out of exigency in the Spanish-American war, in the 1890's, when America supported Cuban liberation from Spain. US forces in the Caribbean destroyed the Spanish fleet and America occupied Havana; this war propelled America into an age of supremacy in the military and geopolitical landscape, as well as medical and scientific primacy.

However, an American troop camp in Cuba was struck by yellow fever in 1900, and Walter Reed was called in. Reed was a Microbiology professor at the Army's medical college; he had also been the one to examine a malaria epidemic in the Washington D.C. and Virginia barracks. At the time, the common notion was that malaria was the result of bad drinking water. Reed however, proved that the cause was not water but something in the air of the swamp. Thus, he was appointed as the head of the yellow fever research team at the time when it was thought that mosquitoes were linked to the distribution of the sickness.

Carlos Finlay, a leading epidemiologist, became a useful aide to the team. Reed conducted tests on volunteering soldiers, which helped to prove that the female mosquito gains the contagious agent when they drink the blood from an ill individual; she was not affected by the illness but would spread the contagious agent when she bit a healthy individual.

Another way to transmit the disease was to inject a healthy human being with the blood of a sick person; they also showed that the sick individual's blood kept the contagion after it was passed through the Chamberland filter. Yellow fever was, thus, found to be a viral infection, not a bacterial one as was thought at first. A year after they began their research, Havana was rid of yellow fever, thanks to Reed's ingenious and methodical work.

The agent that caused yellow fever was isolated in 1927. Rabies, polio and other similar diseases turned out to be caused by viruses, as well; AIDS is another of these diseases. Presently, we know that a virus can also cause tumors and cancers. Cervical cancer is a good example, caused by the HPV virus, which vaccination can help to avert.

THE GOOD VIRUS

Viruses infect bacteria as well. A large number of people think that a virus can only be something that causes a disease, but actually, many viruses benefit the bionetwork of the planet. Bacteriophages are viruses that infect bacteria and eradicate the ones that could be dangerous; many of these bacteriophages are located in or on the human body – in essence, they protect us. A virome is the collective viral group in and on the body, and every person has a different virome, with many common features. More or less half the human population has crassphage, a virus that targets the universal bacteria in the gut bacteroids.

Vaccinations are made from viruses, and viruses are also used for gene therapy, treatment to ensure the transfer of valuable genes into cells; the viruses perform as carriers, to convey the genes that are needed to the cell that needs them. Enzymes produced by viruses are used in molecular biology efforts. Appreciating the benefits of viruses is still an innovative concept. But are viruses alive?

WAIT, WHAT IS LIFE?

Do we see giving birth as an essential element, and, if we do, what about a bacterium that splits into two? Is that also considered giving birth? And a mule that is sterile; does being sterile mean the mule is not alive? Death, as well! Is it essential for something to die in order for it to be considered alive; what, then, do we call the California Bristlecone pines that have been around for ages? If growth is an essential, and if we see absorbing energy as a measure, then we must include the forest fire, which absorbs energy and grows, in the pot. Biology is the science of life; however, biologists still struggle to define it.

SO, VIRUSES?

What are viruses, then? Can we describe them as living organisms? Viruses do result in deadly illnesses. More or less five thousand varieties of them have been extensively researched, but there are undoubtedly millions more. Viruses are made up of DNA or

RNA, but can only flourish in a cell of a living organism. By themselves, outside of a suitable host, they are inert, but If they are introduced into a cell of a host, they annex the mechanics of that cell and start replicating. If they are not in a host's cell, their genetic DNA or RNA is surrounded in a protein or fat cloak, where they can be inactive for a very extended period of time.

They have elements that suggest they are alive, but they do not have all the elements that classify life, such as the cellular mechanics needed to take in nourishment, the capability to live and duplicate on their own, or the energy production needed for metabolic activity. They only have the genetic substance that allows them to annex a host's cell and then do all these tasks.

Austrian Physicist Nobel Laureate for quantum physics, Erwin Schrodinger published a novel about life, in which he describes a living substance as something that has the ability to evade deterioration into disorder and equilibrium; this is modeled after the second law of thermodynamics. In easier terms, it can be described by the marvel that, as long as we are alive, we sustain the form we are in, but when we die, we deteriorate and return to the earth and become a part of it; thus, life can be described as the capability to resist deterioration. Things that are alive do not decay; we maintain ourselves by taking in nutrients, turning them into energy and eradicating unwanted products – what we call metabolism, so that we do not deteriorate.

LIFE ON MARS

In 1976, Viking Lander – NASA's survey spacecraft – landed on the red planet and began tests to see if there were any signs of life. Their studies were aimed at finding proof of metabolic activities. The experiment included adding radioactive fluid nutrients to the soil of Mars with the theory that, if there were life on Mars, it would use the nutrients and discharge radioactive waste products. Not long after the radioactive nutrients were put into the soil, the ship's instruments noticed a carbon dioxide increase. Surprisingly the carbon dioxide was radioactive. This was the cause of the first accounts that there was indeed life on Mars. But later, scientists found that Martian soil has a different chemistry, and can cause responses that resemble metabolism, so this was not evidence that there is life on the red planet.

One of the NASA team members, Benton Clark, registered a minimum of 102 noticeable assets of living organisms. He chose three from this extended list and formed a common description of life:

> *"Life reproduces, life uses energy and it follows a set of instructions embedded within the organism."*

This description is applicable to both DNA and RNA, while lifeless things follow Nature's law, and not any implanted guide within them.

IN CONCLUSION

Viruses do not fall under any specific standards – they have either RNA or DNA; they can duplicate, but they do not use energy autonomously. Instead, they successfully use the energy of the host cell they have annexed. They are stuck in the middle between living and nonliving things. The debate as to where they originate from is ongoing. Were they the first to develop or are they merely parts of cells that were cast off, and then taught themselves to live outside of their former body?

Ed Rybicki, a professor of microbiology from Cape Town in South Africa, has characterized them as forms on the 'Edge of life': *"They are the aliens or orphans in the rich kingdom of life."*

SOURCES AND REFERENCES – CHAPTER 26

Zimmer, Carl. *A planet of viruses*. University of Chicago Press, 2015.

CHAPTER 27:
QUEST FOR SUPERMAN: THE GOOD, BAD AND THE UGLY

"It is better for all the world if, instead of waiting to execute degenerate offspring for a crime or to let them starve for their imbecility, society can prevent those who are manifestly unfit for continuing their kind. The principle that sustains compulsory vaccination is broad enough to cover cutting the fallopian tubes. Three generations of imbeciles are enough."

- Justice Oliver Wendell Holmes

This is an utterly outrageous opinion in the Supreme Court's history. Oliver Wendell Holmes, the Supreme Court Justice at the time, and seen as the most renowned Justice, according to Chief Justice Howard Taft, wrote the seven-to-one view in the case of Buck versus Bell, delivered on the 2nd of May 1927.

THE CASE

The laws of the state of Virginia allowed it to surgically castrate people, without their approval, who were considered psychologically faulty. The laws of the state then also allowed the neutering of those with mental illnesses, epilepsy or anyone considered mentally ailing.

Carry Buck had been raped, and because of this became pregnant. She was considered to be immoral and was institutionalized. At the institution, the doctors made a diagnosis in which Buck was found to be *weak minded.* As a result of this diagnosis, she was sentenced to sterilization, but was simply told she needed an appendectomy, and only understood later what this meant. In 1980, she figured out why she could not conceive after her first pregnancy.

EUGENICS

In the illustrious past of the United States, from a viewpoint that includes democracy, science, economy and military activities, there are some faults: this includes the oppression and ruin of Native Americans, but one that is not very famous to today's Americans is the US Eugenics program.

Charles Davenport, a prominent biologist, began the program in the 1900's after the primary testing thereof, which was done in England. He contributed to the beginning of the Eugenics archive, which was situated on Long Island in New York; at Cold Springs Harbor, to be exact. The goals of the program were to advance the racial stock of

America to be the "fittest" focusing on bodily, psychological and sexual habits of human beings, based on heredity.

They gathered information from families that had unwanted features, such as psychological illnesses, dwarfism, promiscuity, and criminality or from families who had been poor for generations. Scholars and the general public were included in the spectacle. Fairs were held, contests had displays about a 'Fitter Family' or a 'Better Baby.' The Eugenics program was even advertised in movies and books.

An estimate of over sixty thousand people were neutered by this law – even children as young as ten years old, merely for not getting along with their peers. California was the leading state, where over a third of neuterings took place. The Rockefeller Foundation, as well as scientists from Harvard, Stanford, Yale and Princeton supported the program, and gave perverse information that reinforced the program's hypothesis. After the Second World War, when the Nazi atrocities came to light, the laws lost their popularity; but vestiges of them were still found in a couple of states not too long ago – the last of these neuterings was carried out in 1981.

EUGENICS IN ENGLAND

Charles Darwin's cousin, Francis Galton, came up with the concept of Eugenics in England – the word is derived from the Greek term for good birth. Galton studied the aptitude of England's high-class society and determined that it was, in fact, hereditary. He suggested that preferred traits were distributed through family lines. In his publication, 'Hereditary Genius,' he promoted discerning breeding for people. He claimed that, as breeding horses and dogs with the wanted traits was possible, discerning matrimonies could also lead to children with improved aptitudes. The US Eugenics program used this same idea; however, it accentuated eradicating unwanted qualities through neutering.

THE NAZIS' EUGENICS

Adolph Hitler stated to his associates:
> "I have studied with great interest the laws of several American States concerning prevention of reproduction among people whose progeny would, in all probability, be of no value or be injurious to the racial stock."

Madison Grant, an American Zoologist and lawyer, became famous for writing a blatantly racist book, 'The Passing of the Great Race,' and stated that it was his Bible. In the book, Grant describes a 'master race' of Nordic people that were blond with blue eyes. He played a prominent role in attempting to limit immigration from non-Nordic racial stock. This was a time when immigration from eastern and southern Europe was on the rise in United States. The Nazis got their idea from the United States, especially California; eugenics was a pseudoscience, made to eradicate men and women who were seen as unwanted, ultimately to uphold only the Nordic type. Hitler openly altered Grant's Nordic race to a Germanic or Aryan one, arguing to the Germans that their race was frail because of genetic weakening, from breeding with lesser races; the German or Aryan race must, therefore, be cleansed.

Hitler used Darwin's theory of 'Survival of the fittest' – commonly known as Darwinism – as a theoretical justification. This seemed a dream to the German population, as the First World War had devastated them and reduced the economy completely. In 1933, a mandatory neutering law was passed in Germany for people with genetic maladies. In the US, major organizations and institutions, like Carnegie and the Rockefeller Foundation, supported the Eugenic laws. The German neutering movement grew quickly from 1934, with more than five thousand neuterings, to the gassings of the holocaust. The executive secretary of the US Eugenics society, Leon Whitney, commented that, *"While we are pussy-footing around.... Germans are calling a spade a spade."*

Through the power of propaganda, Hitler tried to legitimize anti-Semitism; he crammed it into the pseudoscience of Eugenics to make it more pleasant. The help of the medical community is what made the Nazi doctrine of racial purity and selective survival a success. German doctors performed multiple tests between 1941 and the end of the war on people in the concentration camps. Various methods of neutering and euthanasia were practiced – ranging from large amounts of radiation to electric shock treatment to starvation.

Twins were the most targeted for these studies – one would be put through treatments, infected or hurt, while the other was used for tissue samples or simply killed to use as a cadaver; even their parents were studied to find out why they had twins, in the hopes of advancing the fertility of purebred German or Aryan races. The utmost gruesome and harsh tests were done on people who were still alive; the maniacal enthusiasm of the Nazis to eradicate races they deemed inferior led to the deaths of around six million Jews, 1.8 million Poles, two hundred thousand Gypsies, two hundred and fifty-thousand disabled people, seventy thousand convicts, and many other people who were from opposing political parties.

When the war ended, American Judges presided in 1947 over the Nuremburg trials of the doctors who were accused of performing these torturous and deadly tests in the camps. Two American doctors who were part of the legal group assisted in making the document known as the Nuremberg Code. It has since become the most important document in the records on the ethics of medical research. From it came ten values to protect human subjects from medical study – going from agreement to contribution. After the war, Eugenics became known as a crime against humanity.

BERG'S RESEARCH

When genes, genetics, DNA and the use of DNA studies increased during the 1960's and 70's, Paul Berg - a renowned scientist, did not ignore the idea that the data could be used for the wrong purpose. Berg had been awarded the Nobel Prize for his feat in creating a combined, or hybrid, DNA molecule from two dissimilar organisms - recombinant DNA. Berg, along with other scientists, wrote a document asking that recombinant DNA studies be put on hold until the risks of the studies could be evaluated. In the early 1970's, the risks of mixing dissimilar species' DNA were unidentified. Was it possible that deadlier bacteria could be created? Or perhaps it would create a different kind of cancer or illness. These questions remained unanswered.

Berg headed a board, in 1974, which petitioned the President of the National Academy of Science of the United States to create a voluntary suspension of specific DNA studies that could possibly be dangerous. The academy consists of a group of foremost scientists, and with the help of the National Academies of Engineering and

Medicine, it gives impartial, scientific assistance on serious concerns affecting the nation. Dynamic discussions were held between scientists hailing from across the globe, concerning the facts of the suspension. Papers had a lot of 'what if' situations to cover. Eventually, everyone upheld the suspension.

Berg's committee suggested that a global meeting be held to evaluate the dangers, but that the meeting should not only include scientists but also experts from other fields, so that they could accurately evaluate the dangers and find ways to lessen them. Asilomar, California, was the site of the meeting that Berg and four associates organized in 1975. The purpose of the meeting was for the scientists who suspended their studies to sensibly consider and create the criteria for planning tests for their work that kept in mind the greatest benefits for humanity and circumvented the possible harm to society.

It was a grand success. The International Congress on Recombinant DNA molecules was to be the first topic under discussion, in February 1975. Attorneys, the press and government bureaucrats were also invited, which hailed a new beginning for a new generation in science and the public debate of scientific policies. The meeting laid out strict rules and established approximations of danger for multiple forms of studies, as well as safety necessities appropriate to the danger level.

In July of 1976, the US Government set out rules for DNA study, built on the conference's work, which were very effective. In the more than forty years since these rules were put in place, multiple DNA experiments have been conducted across the globe, with no danger posed to human subjects. The inclusion of the public and other participants created a novel platform for people to see the worries of scientists for the protection of the public in this innovative study into the foundation of life. It established a feeling of safety, knowing that there are precautions. Later, Berg pondered whether a meeting such as the one held in Asilomar, could perhaps solve conflicts across the globe.

The United States funded an epic plan in 1990, to record the complete arrangement of the chemical makeup of the genes in a human cell. The human genome project was intended to be a 15-year task, and estimated to cost around three billion dollars. It proved a groundbreaking effort in understanding the mysteries of the directives of life, and would give insights into illnesses - meaning it could lead to the understanding of life at a molecular level in every way. It was a huge endeavor, because a human cell has three billion base pairs that had to be recognized in their correct order – like finding every word to describe the guidelines of life. The plan was a breakthrough occasion for science and allocated 3% of the expense to research the ethical, legal and social effects of the data gained by understanding genes and genetics.

James Watson, who is known for his work with Francis Crick in coming up with the chemical structure of DNA, was the project lead. He, as well as Senator Al Gore, was listed with the advocates for including the exceptional feature of ethical, legal and social consequences of the project, which was eventually accepted. They were in charge of studying the public effects and creating endorsements to defend society. Their suggestions have created laws that allow discretion of gene data. It bars discrimination based on genetic information in the insurance industry, and workplace.

IN CONCLUSION

As one can see the potential for misuse of such a powerful technology. It is imperative that society remains vigilant in monitoring and protecting human rights and dignity in an environment of profit-motivated business enterprises and rogue governments.

SOURCES & REFERENCES – CHAPTER 27

Kevles, Daniel J. *In the name of eugenics: Genetics and the uses of human heredity.* No. 95. Harvard University Press, 1995.

Summary Statement of the Asilomar Conference on Recombinant DNA. [Online] Available at: https://pnas.org/content/72/6/1981

CHAPTER 28:
MINDS SHAPE THE FUTURE

Our moral responsibility is not to stop the future, but to shape it.

- *Alvin Toffler.*

In life, the only thing we can truly be certain of is death; ageing brings us ever closer to our final day and, as we grow older, we become more vulnerable to illnesses. Cats are usually alive for around twelve to fifteen years; dogs, on the other hand, usually about eight; horses live more or less twenty years, while elephants hang around to seventy. One of the longest known and recorded lifespans is that of the Bowhead Whale, which can live up to two hundred years.

> *But why is it that the lifespans of these animals differ so greatly? Can it be linked to the food they eat, their environment or is it something in their genes?*

Progeria, from the Greek word '*pro*,' meaning *early*, and '*geras*' meaning *old age*, is an uncommon illness that occurs in one out of every eight million babies born. Those affected usually do not live much longer than their teens or early twenties. At birth, these babies seem healthy and normal, but around one or two years of age, they begin displaying signs of rapid ageing. They become bald or they get wrinkles - sometimes they develop illnesses, such as heart disease, that are usually only common among the elderly. *Progeria* is the result of a freak mutation that happens in the first stages of an embryo's growth; it is not transferred to the following generation, so only a couple of uncommon cases occur. A gene mutation is the reason why these people's biological ageing clock is fast tracked.

SWEET 16 AND NEVER BEEN HEALTHIER

The Albert Einstein College of Medicine, situated in New York, has been examining more than five hundred (500) vigorous senior citizens, whose ages range from 95 to 112, as well as their children, in the hopes of finding a gene or genes that is responsible for these individuals enjoying a long and healthy life. The director of the program, Dr. Barzilai, has dubbed the subjects of the scientific endeavor, 'The Super Agers." They have, more or less, all been unharmed by heart illnesses, cancer, diabetes or dementia.

> *"Could Super Agers possess a gene that allowed them to overcome these illnesses, and if it is only a gene, why not study it?"*

However, since we cannot use humans as the subjects of this experiment, and progeria is too uncommon an illness, what do we do now? Instead, a model organism is required that is easy to breed in a laboratory: their maintenance must be inexpensive, and they must share a whole host of genes found in humans. Nobel Laureate, Sidney Brenner, found that effective study in molecular biology was dependent on such a model organism; thus, he presented C. elegans – a 1 mm, dirt-dwelling, nonparasitic round worm. The worm has 1,000 cells, instead of the more than 100 trillion humans have, lives for around twenty-one days, and has many commonalities with the human genome. C.elegans has biochemical trails inside its cells that are similar to humans, as well as between 60 and 80% of similarities in the genes. On this basis, scientists began changing the worm's genes and found something extraordinary; a gene mutation named 'daf-2' was found to multiply the lifespan of C.elegans by two. Another astounding find was that the worm did not just survive twice as long, but it was healthier and livelier for a longer period. Scientists nicknamed the gene the "sweet 16" gene.

The University of California, led by San Francisco's geneticist and biochemist, Dr. Cynthia Kenyon, has been researching this round worm while comparing it with humans. She considers that growing old is not just the body decomposing because of use, but that genes essentially regulate this. Multiple genes have been found that are being thoroughly studied; the mutation that caused longer life in C.elegans was found in flies and mice as well. The possibility that it exists in humans becomes more believable when looking at the data on Super Agers. A counterpart of this mutation, with similar functions, is also found in humans, dubbed "the human daf-2". The daf-2 gene regulates receptors for the *insulin*-like growth factor, *IGF 1*; when these receptors are triggered, growth accelerates. But growth is decelerated by mutated or inefficient daf-2. This slowing down seemingly allows cells to work at a gentler pace, keeping them vivacious and healthy longer than normal. It also appears to resist bacterial infections.

Humans have three types of daf-2 genes, one being the FOXO family of genes. These genes are there to control stress resistance, halting the reproduction and development of cells, as well as apoptosis. Apoptosis, also known as programmed cell death, is a course found in multicellular organisms that terminates cells that are not needed anymore or become a risk to the body, through a mechanism that is similar to controlled suicide. The second family of these genes is known as TOR, which seems to have a role in the ageing process, as well as illnesses that are linked to age. This gene can be blocked by the antibiotic, Rapamycin; administering this antibiotic can prolong the life of old mice. However, it is quite poisonous to the human system, so cannot be used to lengthen the human lifespan.

Venture capitalists, in collaboration with Dr. Kenyon and many other scientists, have begun businesses to create anti-ageing medicines. These enterprises are exploring and attempting to develop medicines that could moderate the properties of "long life" or longevity genes to help mankind. The amount of funding they have generated makes the fountain of youth *that Ponce de Leon* searched for feel more attainable.

Longevity is a product of two factors; nature and nurture, meaning genes and life-style, like diet, exercise, weight, habits like smoking and alcohol intake, etc. When people's DNA is analyzed, it is 99.9% identical in all humans. However, the remaining 0.1% leaves plenty of room for variation. This variation can be used to study the differences in various ethnic and racial groups and even between males and females. The reason for such differences is not related to race, but because of mating patterns. Endogamy includes marrying in the same ethnicity or religious groups, or large-scale mating of a conquering army's males with females of the lands they conquered. Endogamy is a social, rather than a biological process; nevertheless, it creates a distinct group suitable for scientific studies. Analysis of this type of data reveals, for example, that women live longer than men, and African-Americans who survive to the age of 85 live longer than Caucasians in the US. At Hawaii's Kuakini Medical Center, researchers have been following a group of Japanese-American men since the 1960s. They have twice the likelihood of surviving to the age of 100 years. Guess what: they have a mutation in the FOXO gene.

Professor Spencer at King's College London, the same institution which played a key role in taking the best X-ray pictures of DNA, which helped Watson and Crick develop the model of the structure of DNA, has been studying a group of identical twins for over twenty years. Interestingly, the twins share several common features, like height, but the cause of death is frequently different. They have noted striking differences in the incidence of common ailments: only 30% similarity in the chances of developing diabetes, 15% rheumatoid arthritis and 25% age at death. The reason for this marked variation in individuals with exactly the same genetic make-up is what scientists call phenotypic variation with the same genotypic DNA. This is because of factors related to lifestyle, environmental factors like habits, exposure to natural phenomena, like famines, or other outside factors beyond an individual's control. This, at the molecular level, is because of the way a gene's activity is modified as to when it is activated, how much and long it is active, etc. This process leads to different features; diseases in people with the same genetic make-up are called epigenetics. This is a fertile area for research to understand the mechanisms causing these changes. Studies on epigenetics can lead to the development of recommendations for modifying lifestyle, environment and of medicines to improve our health.

Dr. Jaime-Guevera Aguirre is an Ecuadorian endocrinologist, who received postgraduate training in the US and, later, returned to practice in his homeland. He is from Loja, a province in southern Ecuador. Once, travelling by horse in a remote area of Loja, he noted a tribe of dwarfs. Being an endocrinologist, he was intrigued and started studying them. He noted that diabetes and cancer were hardly seen in these people, even though they were obese.

When he first discussed his observations, the obvious objection was that one had to watch over a longer period of time to be sure that it was a true finding. An Israeli endocrinologist, Laron, trained at Massachusetts General hospital, had earlier noted similar findings in a few patients who had migrated from Yemen. Hence this disorder was named Laron's syndrome. Now, it has been over 20 years and, amazingly, these two common diseases, diabetes and cancer, the scourge of other people today, hardly seem to touch the Laron syndrome patients. Genetic analysis of these patients has revealed a defect in the receptor which handles growth hormone and IGF 1. Even though they have adequate production of growth hormone, they are unable to utilize it, because the protein to which it binds on the surface of a cell, to make the cell grow. is defective and does not respond to it. Understanding the mechanisms protecting these patients from diabetes and cancer would be of immense help to the rest of humanity.

VIOLENCE AS A RESEARCH SUBJECT

1997 saw the publishing of *'America's Children: Key National Indicators of Well-Being,'* which was published by the US Federal Interagency Forum on Child and Family Statistics. The publication stated that 2.6 million young people between the ages of twelve and seventeen fell victim to violent crimes in 1994. The US has higher rates of murder and violent crimes than any other industrialized country around the globe.

Biomedical researchers, psychologists and sociologists have been researching the reasons behind this repugnant human conduct.

The question arose to whether or not it was possible to study and evaluate this kind of conduct, as we could an illness. Would there be any results from brain scans or analyzing the chemicals in the bodies of criminals? Studies of depression and anxiety have given valuable results, which led to creating medicines for these illnesses. A large number of violent crimes in industrialized countries are carried out by very few individuals; most of the time recurrent lawbreakers. Is it possible that a gene anomaly is the cause of these crimes, or are these people the result of society's influence?

MONOAMINE OXIDASE

It was 2006 in rural Tennessee; Bradley Waldroup had a little too much to drink and decided to go pay a visit to his estranged wife. He grabbed his hunting rifle and set out to find her. Meanwhile, his wife was dropping off her four children with Leslie Bradshaw, one of her friends, when Waldroup pulled up. He started an argument with the two women, which led to him shooting Leslie, after which he cut open her head. He then went after his wife with a knife and machete; luckily she escaped with her life, although losing her little finger. Waldroup was taken into custody and confessed to the crime. As a result, he faced the death penalty.

But his lawyer took a sample of his blood and sent it to the Molecular Genetics Lab at Vanderbilt University – a rather strange thing to do. Waldroup's defense team asked that they search for a particular gene mutation on Waldroup's X-chromosome – the part that regulates the enzyme known as monoamine oxidase A, or MAO-A. Lo and behold, they found it. MAO-A is in charge of splitting the chemicals, such as dopamine or serotonin, in the brain that control conduct. If there are too many of these chemicals present in the brain, which is possible when this gene is abnormal, it can lead to a loss of control, rash conduct, viciousness and anger. This behavior is particularly magnified when a person is drunk. The jury ruled against the death penalty because of this evidence, instead condemning Waldroup to life in jail.

Waldroup's story is not the first of its kind. The search for a genetic reason behind violence began in Holland in 1978, when a woman asked the University Hospital in Nijmegen for help with her brothers and son, who were suffering from what she thought was some kind of disease or mental illness. The men had been involved in arson, tried to rape someone, and even attempted murder. A thorough family history was conducted, with doctors finding that the violence in the family went back to the 1870's.

Ten years later the research team at the University of Nijmegen found an anomaly in the men – they all possessed the MAO-A mutation on their X chromosome. Men have only one X chromosome; this means the chances of them having the effects of MAO-A mutation occur is much larger than in women, who have two X chromosomes, such that the normal one reduces the effects of the mutation in the other. Women can still pass the mutation on to their children, though.

Scientists used nine-hundred imprisoned violent criminals in Finland as case-studies for their research program, and found MAO-A, as well as a different genetic anomaly known as CDH13, common among them. Interestingly enough, the chances that genes were abnormal or mutated in recurrent offenders were thirteen times larger. Yet, most people with these genes do not become violent criminals: The findings gave rise to even more questions:

Could there be other genes involved as well? Or is violence in some humans caused by other elements, such as alcohol or drugs? Is it possible that psychological childhood trauma or social and economic pressure influences irregular genes, leading to violent tendencies?

A thorough study of twenty-four well-planned programs on the connection of genetic irregularities to violent conduct proposes that 50% of violent conduct is the result of genetic factors. So, that which we receive from our parents and the rest which we are open to in life – stress, diet, drugs, sleep, and so on – ultimately governs our conduct.

HOW ABOUT CLONING OUR ORGANS OR OUR WHOLE BODY AS SPARE PARTS?

The University of Edinburgh had the foresight to establish an Institute of Animal Genetics in 1919, barely a decade after "genetics" in the modern sense was recognized as a discipline. It later grew to become The Roslin Institute. It is a world-class research facility focused on the study of animal biology, with the ultimate goal of enhancing the lives of animals and humans. In 1996, Ian Wilmut, Keith Campbell and colleagues stunned the world by showing a picture of a cloned sheep. This was the first mammal produced by this technology. It was created from cells from the udder of a 6-year-old white sheep. A clone has the same genetic make-up as the cell or body from which it is derived, unlike a baby born from the union of a male and a female, which has a mixture of genes from both.

The lamb was named "Dolly" because it was made from the cells of the udder (breast), and in the words of the scientists who created it, there was no better example of breast tissue than Dolly Parton. Cells from the udder of a white ewe, which were fully developed, mature cells not normally involved in reproduction, were removed. The nucleus, the center of the cell, which has the DNA, was extracted from an unfertilized egg cell or ovum from a black ewe. The nucleus from the white sheep was fused with the cell of the black sheep, whose nucleus had been removed, by using electrical pulses. It was grown in culture for a week and then implanted in a surrogate mother. This was a clone of the white sheep because the entire DNA came from the white ewe. Dolly, the clone was white, grew, mated and produced normal offspring. Sheep normally live to about 12 years, but Dolly was euthanized at six and a half years of age, as she had developed arthritis and a lung tumor.

Cloning not only means cloning a whole animal but, technically, it means producing a genetically identical copy of a biological entity, like genes, cells or tissues. In nature, single-cell organisms like bacteria normally reproduce by splitting into two, having the same genetic material, but in higher species, reproduction is by the fusion of male and female gametes. Identical twins are clones which arise when a fertilized egg splits into two or more and each one develops as an independent baby. Cloning of genes, cells and tissues is being extensively studied to understand the workings of cells, to grow cells and tissues to repair and replace diseased and damaged parts of the human body.

Many animals have been cloned, including cats, deer, dogs, horses and rabbits, to name a few. The cloned animals can be a source of meat, dairy, etc. Cloned embryonic stem cells can be used as replacement cells and tissues. They are very useful in research as they are genetically identical. Cloning can save endangered species and may even recreate extinct species from residual DNA of those organisms.

Could this technology lead to rich people cloning themselves? Or could parents or those who have a loved one terminally ill or dead clone their child or their loved one? Cloning of humans is prohibited in most countries of the world. It is also technically difficult. However, there are companies offering cloning of pets like cats and dogs. Viagen Pets currently charges $50,000 to clone a dog.

CAN SCIENCE FIX DEFECTIVE GENES?

In 2004, the Broad Institute began in Cambridge, Massachusetts, with a partnership between Harvard and Massachusetts Institute of Technology. The goal was to tap the new knowledge from genomic sciences to advance our understanding of life and harness it for treating disease. It was to lay the cornerstone of a new generation of therapeutics, the main goal being to decode the complex parts of life – to find the central molecular root of illness and advance beneficial interventions based on this sound knowledge.

Eli and Edyth Broad funded the project by supplying four hundred million dollars, in the hope of enhancing the quality and quantity of human life. Eli, the son of a Lithuanian immigrant, had a rather difficult time early in his life. To support himself, he sold women's shoes when he was at college. Later, he became very successful financially, and became the only person to create two Fortune-500 businesses.

The United States, more than any other country in the world, has been a benefactor from private donations to science and education. The generosity of the people of the US to science, schools and research exceeded seven billion dollars in 2015.

THE FUTURE OF DNA AND SCIENCE LOOMS BRIGHT

A novel bacterial immune system element was found in 1993. This system, unlike restriction enzymes or molecular scissors, which cut the viral DNA or RNA at specific sequences, is able to recognize a virus which had previously attacked the bacteria, much like an antibody in humans and cuts it into pieces, rendering it harmless. It consists of two components; one to recognize the virus and the second to cut it. This is known as the CRISPR-Cas system. Many groups across the world spent twenty years working on this to understand, manipulate and adapt it to recognize faulty genes in cells. This brilliant and mind-boggling research has produced a system which can now not only recognize a defective gene in a cell but also replace it with a normal one.

A young man named Feng Zhang – only in his 30's – is a lecturer at the Broad Institute; he is seen as a truly excellent mind and a likely candidate for a Nobel Prize. Zhang found a way to mark any singular place in the genome of a human cell during an extraordinary trial in 2013. He could cut the DNA at that spot and implant a different segment of DNA there. In other words, he could pick the abnormal gene amidst the nearly 20,000 genes in a human cell, cut it and replace it with a normal healthy one.

This is gene editing; a magic trick perhaps, or something similar to a spellchecker, that finds the mistake and fixes it.

Zhang furthered the method in 2015, making it even simpler, more effective and less costly. The technology, capable of changing the genome, is currently being explored in more than just single-gene illnesses, but in many other medical conditions as well, such as cancer or drug resistance. Zhang's lab has instructed thousands of researchers in the method and is foremost in sharing the information with the world so as to speed up revolutionary findings.

In every thousand babies born around the world, around ten suffer from a genetic-related issue. Hank Gathers, a 6-foot 7 inches stand out US college basketball player, in the 1988-89 season topped in scoring and rebounding. In a semi-final game played on the fourth of March 1990, he was his dominating self until, after a soaring alley-oop, walking back to his side of the court, he collapsed and died. He was 23 years of age. An autopsy revealed a heart muscle disorder called hypertrophic cardiomyopathy. His life was portrayed in the TV movie, "The Final Shot".

This is the most common cause of sudden death in athletes. It is caused by mutations in the genes involved with the muscle of the heart. Several genes have been implicated, including a gene named MYBPC3. In 2017, an international team of researchers reported on their work with human embryos, generated by fertilizing a healthy donated egg with a sperm from a man carrying an abnormal MYBPC3. The embryos were not going to be implanted. They applied the CRISPR-Cas technology and reported success in correcting the abnormal gene with a healthy one. All embryos were destroyed subsequently.

This is the first step in what would be a complex process of studying and perfecting the technology to make it safe, and reproducible. One has to be sure that it does not have unintended consequences of causing other genetic changes. Even when it is ready for prime time, it can only be of use on a fertilized egg. However, it is a change in paradigm that means, even if one or other parent passes a defective gene, the baby has the potential, in hopefully the very near future, of having the genetic anomaly corrected before birth.

CRISPR technology is being used to change the genes of mosquitoes in the hope of lessening the spread of malaria – a project funded by the Bill and Melinda Gates Foundation. CRISPR sterilizes the female mosquito; therefore, since the male does not have a long life, the mating process is limited. Thus, with sterile mosquitoes spreading across Africa, the mosquito population decreases, as does the malaria virus. TATA trust is funding a similar project in India. As a result of Zhang's exceptional work, CRISPR technology can possibly eradicate insect-borne illness in the near future.

IN CONCLUSION

As researchers chip away at the mysteries of super agers, the bad behavior genes, cloning for repair, replacement of diseased parts, and ultimately CRISPR and similar technologies can edit the genome, one can see in the not-so-distant future, Paradise on this very earth. A future shaped by brilliant minds.

SOURCES & REFERENCES - CHAPTER 28

Longevity Genes Project - Albert Einstein College of Medicine. [Online] Available at: https://www.einstein.yu.edu/centers/aging/longevity-genes-project/

"Aging Genes: Genes and Cells that determine the Lifespan of the Nematode C. elegans". *iBiology*. https://www.ibiology.org/development-and-stem-cells/aging-genes/

"About Progeria". *NIH: National Human Genome Research Institute.* [Online] Available at: https://www.genome.gov/Genetic-Disorders/Progeria

Richardson, Sarah. A Violence in the Blood. *Discover Magazine.* [Online] Available at: http://discovermagazine.com/1993/oct/aviolenceinthebl293

Kolata, Gina. *Clone: the road to Dolly and the path ahead*. Penguin Books Ltd, 1997.

Klotzko, Arlene Judith, ed. *The cloning sourcebook*. Oxford University Press, 2003.

Doudna, Jennifer A., and Samuel H. Sternberg. *A crack in creation: Gene editing and the unthinkable power to control evolution*. Houghton Mifflin Harcourt, 2017.

POSTSCRIPT

WOW! WHAT A MARVELOUS JOURNEY

Science is a beautiful gift to humanity: we should not distort it.
A.P.J Abdul Kalam

What fantastic advance has not been made in DNA and genetics? *Are you left as breathless and excited as I am as to the future of Genetics?*

Although we still have a long way to go to further our knowledge, we are finally able to answer some of the questions ancient geneticists asked, as well as answering the enigmatic questions about the nature of life that have persisted for millennia. Today, we, indeed, acknowledge that the information encoded deeply into the cells of our bodies holds the key to liberation from a wide array of diseases, mental issues and viruses.

Scientists from around the world – the young and the young-of-heart, such as Zhang and Dr. Barzilai, are on the brink of discovering and prolonging life, as well as possibly ridding the world of malaria. Scientists, supported as never before, now have the chance to answer the questions plaguing us:

"What is life really made of?" – A whole array of …..
"How do the cells and organs in the body work?" – Through an intricate network.
"How is it that two separate individuals produce a new life form that contains features of both and the generations before them?" – Through their DNA.
"What is it that determines the array of different skin colors in our world?" - ????
"Why are some children born with anomalies?" - ????
"Why does a cat give birth to kittens, and not puppies?" - ????? "How can a few simple chemicals produce such a panorama of life forms on our planet?" - ?????????

Although we have found answers to at least some of the questions above, others – ancient and new - still remain unanswered. Now is the time, therefore, for an aspirating scientist such as you to take on the adventure of advancing the study of genetics to the next level. Only by furthering our knowledge, and avoiding another dark age, can we move forward and ensure a better way for human society. We also need the society to be vigilant in safe guarding against misuse of such powerful tools.

I wish you well, dear reader, and I hope that this book will inspire you to become one of the most famous scientists in the world, and that one day you may also feature in a book such as mine, and much will be written about the phenomenal scientific work you have done.

Fare well, my dear aspiring scientists, and ensure your life is always filled with enough questions to answer, since life is enigmatic and meant for exploration.

With my kindest regards,
Dr. Haq

Human Cell Structure

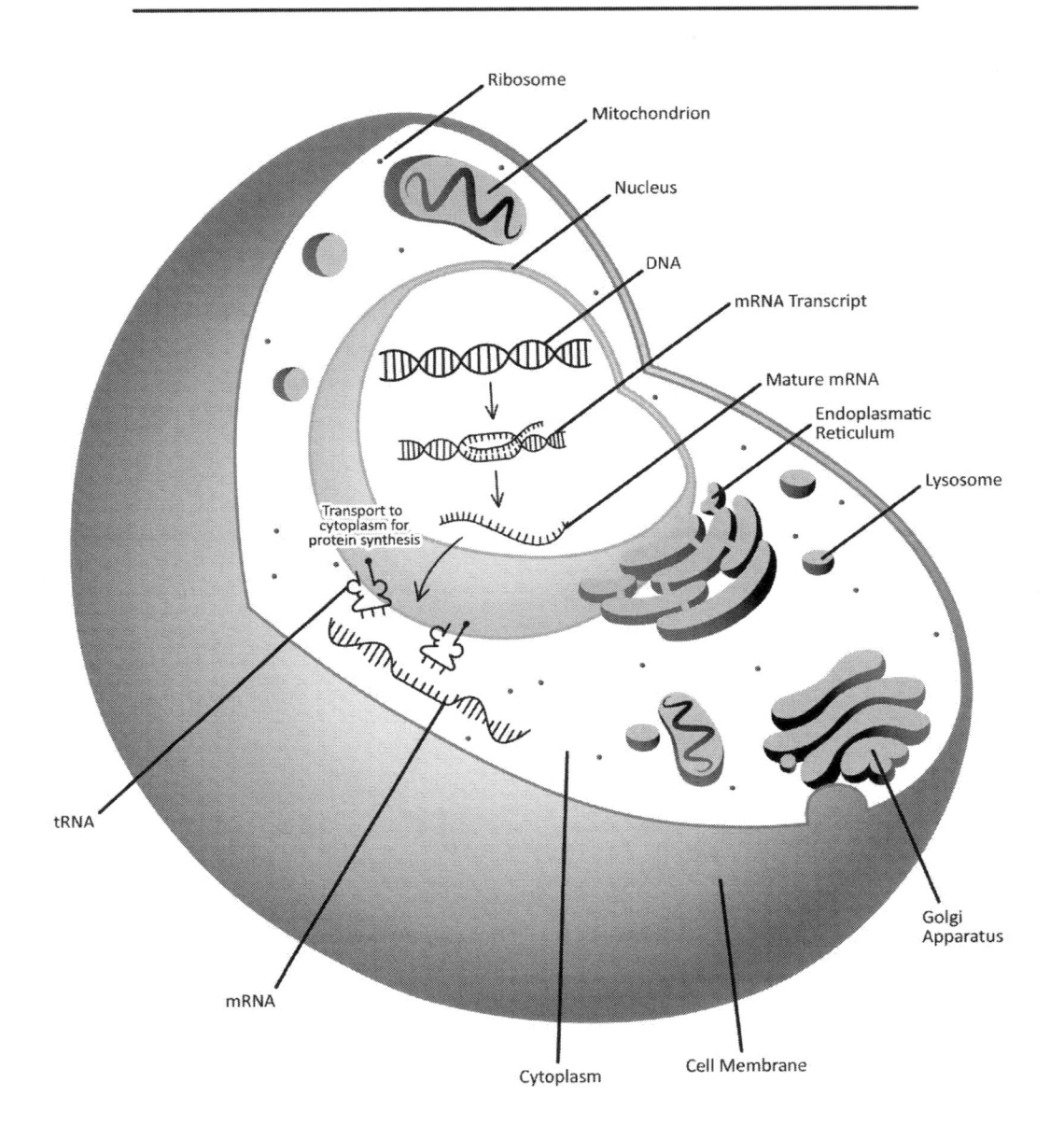

DIAGRAM OF HUMAN CELL STRUCTURE

FURTHER READING

Mukherjee, Siddartha. *The Gene: An intimate history* Scribner, New York, 2016

Watson, James. *The Double Helix: A Personal Account of the Discovery of the Structure of DNA.* Penguin, London, 1969

Berry, Andrew and James Watson. *DNA: The Secret of Life.* Knopf Doubleday, New York, 2003

Judson, Horace F. *The Eighth Day of Creation.* CSHL Press, New York, 1996

Kean, Sam. *The Violinist's Thumb: And other Lost Tales of Love, War, and Genius, as Written by Our Genetic Code.* Little Brown and Co., New York, 2012.

McCabe, Edward and Linda McCabe. *DNA: Promise and Peril.* University of California Press, 2008

Venter, J. Craig. *Life At The Speed Of Light- From the Double Helix to The Dawn of Digital Age.* Penguin, New York. 2013

ABOUT THE AUTHOR

Mohamed M Haq is a medical Oncologist, trained at the nationally renowned MD Anderson Cancer Center. He is on the clinical faculty of Baylor College of Medicine. As a practicing cancer doctor for over thirty years, he has developed and polished the skill of conveying complex medical information to patients ranging from space scientists to cattle ranchers. His scientific review article on this topic in the journal, *Texas Medicine,* was well received and cited for meritorious writing of a scientific review article. He has published scientific articles and articles on history and the history of medicine. In 2009, he wrote a web-based book, *Charity Clinic 101: Guide to Operating a Charity Clinic.*

A

Adam, 23, 24, 169, 170
Adenine, 119, 161
Adolph Hitler, 185
Al Gore, 188
Alec Jeffries, 160
Allison, 149, 150, 153
amino acids, 124, 127, 129, 132, 133, 171, 172, 175
Anaximander, 47
Ancient Egypt, 10
ancient Indian, 17, 18, 21
Anglo Saxons, 58
animalcules, 71, 72, 74
Anthropology, 165
anti-ageing, 192
Antoni Leeuwenhoek., 71
Apoptosis, 192
Arabs, 17, 63
Aristotle, 46, 47, 49, 50, 51, 52, 53, 60, 75, 79, 80, 93
artificial fertilization, 63
Aryan, 44, 185, 186
Asilomar, 187, 188, 189
astrologers, 55
atoms, 6, 48, 78, 119, 124
Avery, 107, 108, 109, 110, 111, 112, 113, 114, 118, 140
Avicenna, 52

B

Babylonians, 63
bacteria, 74, 76, 107, 108, 109, 110, 111, 112, 118, 123, 130, 135, 137, 138, 140, 141, 142, 144, 147, 172, 177, 179, 187, 197, 199
Bacteria, 136, 138, 141
Baghdad, 33, 50
Banting, 145
Beijernick, 178
Benton Clark, 181
Berg, 134, 138, 139, 140, 141, 187, 188
Best, 145
bewitched, 59
Bhagavad Gita, 16
Bible, 22, 23, 24, 26, 28, 29, 33, 61, 78, 185
Bichat, 81
Bill and Melinda Gates Foundation, 201
binomial system, 66, 68, 69
bioengineered, 143
biological Eve, 169
birth, 3, 9, 12, 14, 16, 19, 20, 25, 28, 29, 31, 37, 38, 41, 43, 44, 51, 53, 55, 64, 82, 88, 152, 163, 180, 184, 190, 200, 203
Blackwell, 86
blasphemy, 89
blood relatives',, 50
Boyer, 135, 136, 140, 141, 144
Brahman, 18, 21

Broad Institute, 198, 199
Brunn, 94
Buck versus Bell, 183
Buddhism, 40

C

C. elegans, 191, 201
Calvin Bridges,, 65
Camerarius, 63
Campbell, 197
Canon of Medicine, 52
Carry Buck, 183
cell', 78
Celts, 58
centaurs, 55
Chamberland, 177, 178, 179
Charaka, 20
Chargaff, 113, 118
Charles Darwin, 70, 93, 184
Chimera, 140
Chinese, 37, 38, 39, 40, 41, 42
Christian, 12, 29, 61, 65, 78, 89, 95
Christianity, 22, 59, 61, 158
chromosome, 24, 97, 102, 104, 169, 195, 196
chromosomes, 24, 65, 102, 103, 104, 138, 196
Church, 59, 60, 61, 64, 78, 108
Cicero, 56, 57
cloned sheep, 197
cloning of pets, 198
clot, 154
clotting factors, 147
Cohen, 85, 135, 136, 140, 141
Columbia University, 100, 101, 116, 118
Confucius, 37, 41
consanguinity, 131
Copernicus, 62, 65
creation, 12, 13, 28, 33, 34, 35, 37, 39, 43, 47, 64, 69, 78, 79, 89, 134, 140, 174, 202
Crick, 113, 115, 120, 121, 122, 123, 124, 135, 174, 188, 193
CRICK, 115, 116, 120
CRISPR, 199, 200, 201
Cromwell, 156
cross-pollinated, 97
cross-pollination, 63
cytoplasm, 124, 126
Cytosine, 119, 161

D

daf-2', 191
Dalton, 78
Darwin, 49, 86, 87, 88, 89, 90, 91, 92, 93, 101, 153, 165, 185
Darwinism, 185
Davenport, 184
De Kruif, 138
Democritus, 48

138

deoxyribonucleic acid, 112, 161
deoxy-ribose., 117
Diabetes, 145
dissection, 62, 160
DNA, 3, 5, 6, 14, 36, 38, 99, 106, 107, 110, 111, 112, 113, 114, 115, 116, 117, 118, 119, 120, 121, 122, 123, 124, 125, 126, 128, 129, 130, 135, 136, 137, 138, 139, 140, 142, 144, 145, 147, 156, 159, 160, 161, 162, 163, 165, 167, 168, 169, 172, 174, 180, 181,182, 187, 188, 189, 192, 193, 197, 198, 199, 203, 205
DNA ligase,, 139
DNA test, 159, 160
Dochez, 109
Dolly, 197, 201
dominance, 51, 97
dominant, 27, 97, 98, 152
Dominant, 97, 98
Doppler, 95
Dr. Cynthia Kenyon, 191
Dr. Jaime-Guevera Aguirre, 194
Dr. Johan Rothman, 67
Dr. Kilian Stobaeus,, 67
Drosophilists', 105
Dubos, 110
Dutch, 58, 71, 72, 73, 74, 90, 101, 178
Dutrochet, 82, 83

E

E. coli, 126, 139, 140, 141, 145
E.coli, 126
Edward Tatum, 132
egg, 28, 51, 64, 104, 123, 173, 197, 198, 200
Egyptian, 10, 11, 12, 13, 14, 130, 174, 176
Eli and Edyth Broad, 198
Eli Lilly, 143, 144, 146
Empedocles, 48
ensoulment, 60
enzyme, 110, 133, 134, 137, 139, 140, 195
Epicurus, 48
epigenesis, 51, 64
epigenetics, 193
Eternal life, 40
Eugenics, 184, 186
Europe, 2, 33, 50, 60, 63, 65, 86, 100, 118, 168, 185
Eve, 169, 170
evolution, 5, 54, 68, 69, 70, 89, 90, 91, 93, 101, 105, 170, 174, 202

F

Factor V Leiden, 154
Falkow, 136
father's sperm, 65
female semen, 52
fertilization, 23, 60, 64
finches, 87, 88
Finlay, 179

Fly Room, 100, 104
FOXO, 192, 193
Francis Galton, 184
frog, 135, 142
fruit flies, 100, 101, 103, 106, 132
fuse, 144

G

Galapagos, 87
Galen, 52, 53, 57, 65
Galileo, 65
Gamow, 123, 125
Garbha, 20
Garrod's, 130, 133
gemmules, 49, 90, 93, 101
gene, 24, 90, 97, 98, 101, 104, 105, 106, 107, 118, 131, 132, 133, 134, 135, 137, 138, 140, 141, 142, 144, 151, 152, 153, 155, 161, 174, 179, 189, 190, 191, 192, 193, 195, 199, 200
gene editing, 199
Genentech, 143, 144, 146, 148
genes, 4, 27, 58, 97, 98, 105, 106, 112, 131, 133, 134, 135, 137, 138, 139, 141, 144, 147, 152, 153, 161, 162, 175, 179, 187, 188, 190, 191, 192, 196, 197, 199, 200, 201
genetic fingerprint, 163
genetics, 2, 3, 4, 5, 6, 7, 26, 30, 36, 53, 61, 102, 105, 120, 126, 131, 132, 134, 149, 161, 168, 187, 188, 197, 203, 204
Genetics, 3, 99, 106, 131, 189, 195, 197, 203
genome, 105, 138, 139, 161, 162, 188, 191, 199, 201
genotypes, 97
genotypic, 193
George Beadle, 131, 134
Germans, 58, 185, 186
GMO's, 91
God, 6, 8, 18, 21, 22, 23, 24, 33, 34, 35, 61, 62, 69, 70
gods, 8, 13, 38, 39, 41, 48
Goebel, 110
golden rice, 147
Greek, 29, 43, 46, 47, 48, 49, 51, 52, 53, 55, 56, 57, 59, 60, 64, 65, 75, 78, 108, 140, 184, 190
Griffith, 110, 111
Guanine, 119, 161

H

H.M.S Beagle, 87
Hamilton Smith, 137
Hank Gathers, 200
Heidelberger, 109, 110
helix, 119, 122, 125, 161
hemoglobin, 150, 151, 152
hemophilia, 24, 36, 154
Heraclitus, 47
Herbert Boyer, 135, 143
hereditary, 36, 50, 90, 113, 118, 122, 131, 138, 150, 151, 161, 167, 168, 185

heredity, 5, 7, 23, 26, 31, 38, 44, 49, 53, 59, 61, 62, 90, 96, 97, 98, 99, 101, 102, 110, 116, 120, 131, 184, 189
Hermann Steller, 105
heterozygous, 153
Heyerdahl, 165, 166, 167
hieroglyphics, 8, 15, 16
HINDUISM, 16
Hippocrates, 49, 107
Hoagland Laboratory, 108
Homo sapiens, 69
homozygous, 152, 153, 154, 155
homunculus, 64
hormone, 143, 194
hormones, 147
Hugo de Vries, 90, 101
human insulin, 144
Humulin, 145, 146
Huntington's disease, 152
hybridization, 63, 96
hypertrophic cardiomyopathy, 200

I

Ian Wilmut, 197
Iceman, 168
immune system, 136, 199
immunization, 177
India, 2, 16, 17, 19, 21, 33, 125, 149, 151, 172, 176, 201
Indiana, 143, 151
Indianapolis, 145, 146
infidelity, 59
Ingram, 152
inheritance, 15, 16, 22, 23, 27, 30, 31, 36, 37, 38, 50, 53, 59, 93, 94, 96, 97, 98, 99, 101, 102, 105, 131
Inheritance, 97
inherited diseases,, 19
Innocence Project, 163
inorganics, 173
Inquisition, 59
insulin, 141, 142, 143, 144, 145, 146, 147, 192

J

J.B.S. Haldane, 171
Jean-Baptiste Lamarck, 88
Jefferson, 57
Jeffries, 161, 162, 163
Jenner, 176, 177
Judeo-Christian, 13, 28, 30

K

Khnum, 8, 11, 14
KHORANA, 125
Koelreuter, 63
Konrad Spindler,, 168
Kon-Tiki, 166

L

Lamarck, 90, 96
Laozi, 41
Laron, 194
Laron's syndrome, 194
last universal ancestor, 172
Latin, 56, 74, 75, 105
Laws of Manu, 19
Leonardo Da Vinci, 62
Leucippus, 48
Levene, 116, 117
life on Mars, 181
Linnaeus, 62, 66, 67, 68, 69, 70
LINUS PAULING, 119
living substance, 181
logos, 48
LUA, 172
LUCA, 171, 172

M

Madison Grant, 185
Mahabharata, 17, 18
Maitland, 176
malaria, 149, 150, 153, 155, 178, 200, 203
Malpighi, 82
Malthus, 88
MAO-A, 195
master race', 185
maternal, 19, 50, 167, 168, 169
Mathaei, 126, 127, 128, 129
matrilineal, 38, 58
McCarty, 112
McLeod, 145
Mendel, 61, 63, 93, 94, 95, 96, 97, 98, 99, 100, 101, 131
metabolism, 131, 133, 139, 174, 181
microbiological life, 71
microscope, 50, 64, 71, 72, 73, 74, 76, 81, 82, 102, 103, 110, 151, 160, 178
Miecher, 114, 116
Miescher, 116
Miller, 171, 172, 174
Mirsky, 113
Mitochondria, 167
molecular disease', 152
molecular scissors, 137, 161, 199
monastery, 94, 95
Morgan, 100, 101, 102, 103, 104, 105, 106, 131
MRCA, 169
mutation, 90, 103, 104, 105, 132, 133, 152, 153, 154, 190, 191, 192, 193, 195, 196
mutations, 90, 101, 102, 104, 105, 132, 133, 154, 155, 163, 167, 200
MYBPC3, 200

140

N

National Academies, 63, 187
natural selection, 86, 88, 89, 90, 91, 93, 102, 153, 155, 172
Neurospora, 132, 133
neutered, 184
neutering, 183, 185, 186
NIH, 126, 129
Nirenberg, 3, 126, 128, 129
NIRENBERG, 126
Nobel Prize, 2, 106, 114, 122, 125, 129, 133, 137, 187, 199
Norway, 68, 166
nucleic acid, 113, 115, 117, 118, 161, 173, 175
nucleic acids, 113, 116, 117, 125, 139, 161
Nuclein, 116
nucleotide, 117, 125, 127, 162
nucleus, 83, 103, 116, 124, 126, 167, 197
Nuremberg Code, 186
Nuremburg trials, 186
Nuwa, 39

O

Ochronosis, 130
Oliver Wendell Holmes, 183
On the Origin Of Species, 89
organic, 89, 99, 109, 139, 173
Oswald T. Avery, 107
Otzi, 168

P

pancreas, 143, 144, 145
pangenesis', 90
papyrus, 11
parasite, 74, 149, 150, 153
Pasteur, 76, 80, 177
paternal, 19, 50, 108
patriarchal, 58
Pauling, 121, 152
peas, 96, 97
Pehr Artedi, 68
persecutions, 59
Peruvians, 166
phenotype, 97
phenotypic, 193
photosynthesis, 172
plant hybridization, 63
Plants, 62, 95
plasmid, 139, 140, 141, 142
plasmids, 138, 140, 141
Plato, 45, 46, 52
pneuma', 75
Pneuma', 60
pneumococcal, 108, 109, 110, 112
polio vaccine, 143
pollination, 96
Polynesians, 166, 167

preformation, 51, 64
President Kennedy, 128
primordial soup, 171, 174
Progeria, 190, 201
protein, 119, 121, 124, 125, 127, 128, 129, 133, 134, 145, 151, 152, 173, 174, 180, 194
proteins, 112, 116, 117, 120, 123, 124, 126, 132, 133, 142, 144, 154, 171, 174
pyramids, 10, 12, 16
Pythagoras, 50

R

Ra, 8, 13
recessive, 27, 97, 98, 152
recombinant, 106, 135, 140, 144, 145, 146, 187
Recombinant DNA, 147
Redi, 80
Reed, 178, 179
reincarnation, 40
Remak's, 84
Renaissance, 33, 50, 60, 65, 80
restriction enzymes, 137, 139, 199
restriction length polymorphism, 162
RFLP, 161
ribose', 117
ribosomes, 124, 126, 127, 129
rice, 146, 147
RNA, 3, 124, 125, 126, 127, 128, 129, 133, 171, 173, 174, 180, 181, 182, 199
Robert Hook, 72
Rockefeller Institute, 108, 111, 117
Roman, 50, 52, 53, 56, 57, 58
Rosalind Franklin, 121, 122
Roslin Institute, 197
Royal Society, 71, 74
Rybicki, 182

S

Saint Augustine, 78
San Francisco, 135, 143, 144, 191
Santana Dharma, 18
Scandinavia, 58
Scandinavians, 58
Schleiden, 76, 81, 82, 83
Schrodinger, 120, 181
Schwan, 82, 83
Schwann, 76, 78, 83
scripture, 12, 28
segregation, 97, 98
seven 'Eves, 169
sexual reproduction, 63
Shakespeare, 80
showed, 63, 64, 65, 67, 69, 80, 81, 84, 95, 98, 111, 112, 113, 117, 118, 121, 124, 144, 150, 153, 157, 169, 177
Sickle Cell, 150
sickle cell disease, 151, 152, 153

sickle cell trait, 150
sickle hemoglobin, 150
Sidney Brenner, 191
Smallpox, 176
smallpox vaccine, 177
Socrates,, 45
Solanum Lycopersicum, 66
soul, 13, 14, 16, 18, 20, 21, 23, 29, 33, 34, 35, 36, 39, 40, 41, 51, 60, 62, 173
Soul, 60
souls, 14, 33, 34, 36, 40
sperm, 28, 50, 64, 65, 74, 104, 123, 200
St. Augustine, 60, 79
St. Thomas Aquinas, 60
Stanley N. Cohen, 135
Streptococcus, 107
stromatolite, 172
stromatolites, 172
Sturtevant, 102
SV40, 139
Swanson, 143, 144
Sweden, 68, 69
sweet 16" gene, 191
Sydney Brenner, 124
Sykes, 168, 170

T

Taoism, 41
temples, 10, 12, 13
Tertullian, 60
Thales, 47, 53
The Book of Medicine, 57
The Final Shot, 200
The Super Agers, 191
theological, 68, 90
thrombophilia, 154
thymine, 117, 119
Thymine, 119, 161
toad, 141
TOR, 192
traits, 14, 16, 19, 21, 22, 31, 32, 36, 37, 38, 43, 96, 102, 105, 107, 113, 116, 118, 122, 150, 161, 185
transmigration, 14

U

Ungar, 95
unicorns, 55
universities, 60, 62, 102
University of Lund, 67
Upanishads, 19
Uppsala University, 68

uracil, 127
Urey, 171, 172, 174
uterus, 52, 64
UUU, 127, 128

V

vaccines, 147, 153
variolation, 176
Vedas, 16, 18, 21
Viagen Pets, 198
Virchow, 76, 84, 130
virome, 179
virus, 137, 139, 176, 178, 179, 199, 201
viruses, 120, 137, 179, 180, 203
vital force, 173
Vitalism, 173
vitamin, 132, 133
Vitamin A, 146, 147
vitamins, 132, 133, 146, 147
Vitamins, 146

W

Waldroup, 195
Wallace, 89
Watson, 113, 115, 120, 121, 122, 123, 125, 128, 135, 188, 193, 205
Werner Arber, 137
Wilkins, 120, 121, 122
William Bateson, 131
William Harvey, 64
Wilson, 101

X

X-ray diffraction, 119, 121
X-rays, 120, 133

Y

Y chromosome, 169
Yama-Rajas, 40
yang, 37
Yellow fever, 178, 179
Yellow River, 39
yin, 37

Z

Zeno, 48
Zhang, 199, 201, 203

Made in the USA
Columbia, SC
19 August 2019